GOD FAVORS THE BOLD:

VOICES OF THE TEXAS NAVY
1 8 3 6 - 1 8 4 5

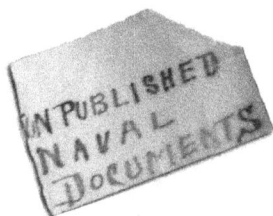

COMPILED & EDITED BY MICHELLE M. HAAS

Copano Bay Press
2013

ISBN: 978-0-9884357-5-9

CONTENTS

PUBLISHER'S NOTE

This is not a history of the Texas navies. This book is intended to be a collection of writings about our navies by the Texians who took part in the adventures. In many cases, the selections contained here have been used or excerpted in other works, but have not been presented in their entirety. In some cases, they haven't seen print since they were originally published in the 19th century.

We, as Texans, have an intense interest in the ground actions leading up to (and of course, including) the Battle of San Jacinto, and rightfully so. We're Texans! We are the people of Sam Houston and Big Foot Wallace, Uncle Jimmie Curtis and Brit Johnson. We remember Goliad and we remember the Alamo, but do we really remember the Battle of the Brazos? The celebration of our heroic heritage still too often omits the men who sailed under the single-starred ensign and menaced the shores of the enemy. That Texas had hundreds of miles of coastline vulnerable to attack and invasion without naval support cannot be denied. That our naval presence in the Gulf diverted men, arms and funds from the Mexican warchest to facilitate protection of her port towns cannot be denied. Our navies served a lofty purpose and their stories, like the stories of every band of men to fight for Texas, are uniquely Texan... underfunded, hastily drawn together groups defying all odds to prove that everyone has the opportunity to do the extraordinary in Texas. So we'd do well to remember that were it not for our small but determined navies, our history as we know it might have been considerably different.

Whatever may be wrong between these covers, I delight in taking full responsibility. I hoarded the transcription, research, annotation, design and editing because I've long had a soft spot for the Texas navies. To the memory of

Samuel Rhoads Fisher and Edwin Ward Moore this humble collection of words is dedicated. Both men sacrificed their careers at the altar of Sam Houston, trying to keep a navy afloat in the Gulf.

-Michelle M. Haas, Managing Editor
Windy Hill

Thumbnail Sketch of the First Texian Navy

The First Texian Navy was borne out of the provisional government's desire to protect the coast, in the final months of 1835. The soon-to-be Republic did not have the wherewithal to formally organize a navy just yet, so letters of marque and reprisal were issued to privateers while a national naval force was gotten up. In the first two months of 1836, four vessels became the First Texian Navy: *Independence*, *Invincible*, *Brutus* and *Liberty*.

The career of the First Navy was brief but ambitious. The *Liberty*, a privateer turned national vessel, was the first ship purchased by the Republic, well before independence had been declared, and made a cruise of the Gulf, capturing the Mexican schooner *Pelicano* in February or March, 1836 (sources differ as to the date of the capture.) In April, the *Invincible* took the U. S. brig *Pocket* as a prize, causing quite a stink that would continue stinking for several years to come. The *Liberty* guarded the ship bearing a wounded Sam Houston after San Jacinto to New Orleans. There, she was to be refitted to continue in her promising career. The Republic couldn't pay the repair bills, so the *Liberty* was seized in May and sold to keep the wolves at bay. During the summer of 1836, Texian vessels blockaded Matamoros, then spent fall and winter in repair in New York and New Orleans.

1837 marked the end of the remaining three vessels of the First Navy. In April of that year, the *Independence*, while conveying William H. Wharton to New Orleans, fell in with the *Vencedor del Alamo* and *Libertador* at the mouth of the Brazos. A running fight up the river ensued, until it reached the vicinity of Velasco, where the *Independence* could go no further. The ship was surrendered as the town looked on, along with S. Rhoads Fisher, Secretary of the Navy.

Secretary Fisher, frustrated with the underfunded state of things, joined the remaining two ships on what would be their final cruise. During the summer of 1837, the *Invincible* and *Brutus* were very busy capturing prizes, claiming Cozumel in the name of Texas, setting fire to enemy settlements and ruling the Gulf. Sam Houston did not approve of Fisher's absence and took steps that would end the Secretary's career. Meanwhile, the newspapers and the general population hailed the return of the victorious navy to Galveston on August 26, 1837.

The *Brutus* was sent into port with Secretary Fisher and a prize in tow. The *Invincible* stood out and waited until the following day to clear the bar. But when the summer sun rose in the morning, two sails were seen chasing the merchant schooner *Sam Houston* toward the bar. They were Mexican brigs and their attention soon focused on the *Invincible*. The *Invincible* put up a good fight and attempted to lure the *Libertador* and *Iturbide* toward shoal water, but to no avail. The Mexican vessels stood off.

Preparations were immediately made upon the *Brutus* to render aid to her fellow vessel. But the bar got the better of both ships. The *Invincible* bogged down on the bar on the evening of the 27th, trying to make a run for the harbor, and was dashed to pieces on the bar overnight. The *Brutus* also ran aground, but was able to be freed and return to Galveston, where she was ruined by a storm on October 5, 1837. All of the ships of the First Texian Navy were dead and the political fireworks between Secretary Fisher and President Sam Houston had begun to fire.

CRUISE REPORT:
FIRST & LAST VOYAGE
OF BRUTUS

To the Honorable Secty. of the Navy

Report on the Cruise & Transactions of the
Texian Schooner of War *Brutus*

Saturday, June 10, 1837, pursuant to orders received, I got underway and stood out to sea in company with the *Invincible*. After having conveyed the Texian schooner *Texas* to Matagorda bar, returned again to Galveston bar, sent a boat on shore which returned at midnight. We immediately got underway and stood to the east, cruising near the mouth of the Mississippi in hopes to fall in with some of the Mexican vessels, but not succeeding, stood for the coast of Mexico. On the 1st of July, parted company with the *Invincible* having previously agreed to rendezvous at the Island of Mujeres.

Cruised some days near Cape Antonio on the coast of Cuba but nothing appearing, run for the Isle of Mujeres on the 7th, made the Island of Contoy and Mujeres on the 8th. Anchored in company with the *Invincible*, in a few days completed wooding and watering the vessels. Made several exertions to the neighboring islands and mainland. Found abundance of turtles in pens and helped ourselves. Caught some small pirogues of but little account; destroyed some. Liberated all prisoners on the 12th.

Stood out to sea again and run down to the Island of Cozumel on the 13th, anchored the vessels on the southwest point of the Island. Landed with our boats and planted the Single Star Banner of our country on the soil of this delightful island. The inhabitants were but few but expressed their good feelings for us at the same time swearing allegiance to our cause. We made such surveys and remarks as our limited time would admit of. The anchorages are indeed safe and commodious for any number of vessels. The soil is delightful; the climate salubrious; the forests abound in the finest kinds of timber—logwood, mahogany and Spanish cedar—and abundance of fruits of various kinds. There is also an abundance of water. On the whole, I think it is a most desirable acquisition to our government and I would respectfully recommend it to the consideration of our congress.

On the 16th of July started again for the Mexican coast. On the 17th anchored on the west side of the Island of Contoy. Found domestic animals but no inhabitants, although there were recent marks of people having been in the houses. Found many pens full of turtles. Took a fresh supply on board. Sent our boats on an expedition to a town said to be near Cape Campeche. Next day, they returned, unable to find the place. Brought a canoe with them having on board nearly or perhaps all the saints in the calendar with some female toggery and a whole host of virgins. Alas, they were all composed of moss.

19th, sailed again for the coast of Yucatan. 21st, landed at Silan but did not find anything of consequence. From thence, stood down to Telchac. I sent Lieut. Wright on shore to take the town. This he soon accomplished, the Alcalde making a formal surrender of the town to the Texian government.

22nd, captured schooner *Union* of Sisal, loaded with logwood; chased several vessels but they were all neutral; captured a number of pirogues, some having valuable property on board. On the 24th, anchored off the town of Chuburna. Here I accompanied Judge Fisher. Landed on the beach unarmed. In a few minutes discovered a squad of cavalry within one hundred yards of us. We lost no time in reaching the boat. We soon had her afloat again and ready for action. As they approached, the Judge drew a pistol and shot one of them. We exchanged fire from pistols. In all, they discharged all their arms. Amongst them were two large pieces that were placed on the ground and deliberate aim taken at us. Not one, however, came within twenty yards of us.

Returned on board again, manned and armed two boats, went on shore and burned two towns. Our gig was absent two days which caused me to think something had happened to her. She, however, returned this evening from Sisal bringing intelligence that there was a Spanish vessel there. We stood down and found the vessel. She was a Spanish merchant schooner without cargo. This day we captured two Mexican schooners, the *Adventure* and *Telegraph* of Campeche. We then anchored abreast of Sisal in company with the *Invincible* and our prizes.

On the 26th, Capt. Thompson sent a canoe on shore with a letter demanding of the Commandant a sum of money to prevent his destroying the town. They, however, did not send off any answer. On the morning of the 27th, hauled close into the town with the flag of truce still flying. At 7 a.m., the castle bearing south, those miscreants fired a shot at our vessel. We took no notice of it for some time until we got close in, then anchored and fired two shots, but finding through the bad quality of our powder they did not reach the fort, got underway again, it being nearly calm, got our boats ahead and towed in. All this time, the enemy kept up a fire from some heavy pieces of cannon. We occasionally returned the compliment. Previous to our firing, we hauled down the white flag and hoisted at our top gallant masthead the national standard.

On approaching as near the shore as we could with safety, being in 13 ft. water, the castle bearing S.S.W., anchored with a spring on the cable and opened fire from our starboard side and pivot gun. The *Invincible* was nearer the beach and kept up a heavy and incessant fire upon the town. No sooner had we commenced our fire than the enemy, ascertaining with precision our position, opened a very heavy fire from the castle, the round fort and large gun on the beach. Finding that we had all to lose and but little to gain, as we had more than one thousand men to contend with, thought it most prudent to shift our quarters further out.

On this day, every officer and man on board done his duty undauntedly and cheerfully. We burned the schooner *Adventure* here, as she was a dull sailer. From this place, after landing our prisoners, forty in number, stood up for the Alacranes Islands. On the 31st, anchored inside. The *Invincible* found the schooner *Albispa* with a cargo on board. 1st August, a vessel hove in sight. I went in chase of her but unfortunately she sailed too fast for us. Returned again along the vessel and anchored in the port.

On the morning of the 3rd of August, saw a strange sail to windward. Immediately got underway and stood out in chase. She carried all sail from us. At 10 a.m. I succeeded in getting alongside of her. Found her to be the British schooner *Eliza Russell* from Liverpool bound for Sisal with a cargo of merchandise, the greater part appertaining to a Mexican merchant in Merida and by a letter directed to him, found that it was not insured but at the sole risk of said merchant. I put an officer and crew on board and took her into Alacranes for the decision of Capt. Thompson who immediately sent her to Galveston for adjudication.

Sunday, the 6th, I went on shore and searched the Islands. Found many articles of various description buried in the sand which we too on board and, after hoisting our flag on shore and taking possession in the name of the government, sailed again on our cruise. Of these islands I would say much but at another time. On the east reefs, we found the wreck of a British vessel and left her undisturbed, she not having either mast, rigging, sails, anchors or cables. Run again for the coast of Yucatan on the 10th of August. Run into Campeche Bay. Got a canoe alongside and put all our prisoners in her, five in number and some belonging to the *Invincible*, and sent them on shore much gratified. Stood down along the coast for Laguna. Staid there one day, but not finding anything proceeded along.

On the morning of the 12th off Tabasco Bar fell in with and captured the Mexican packet schooner *Correo* from Vera Cruz for Tabasco. In overhauling all the letters, found some from the authorities stating that the Mexican squadron were all in Vera Cruz and that the *Independence*, the *Teran* and *Bravo* were to sail to eastward. We cruised three days in the track along the land endeavoring to intercept them, in case they came out, but they did not. Stood into a small place called Chiltapec to land our prisoners and obtain water, but as the boats approached the shore, the officers discovered a body of troops on the beach. They immediately returned on board. Next morning at daylight, hoist our white flag, put all prisoners in the boat and sent them on shore. Amongst them was an old lady, mother in law of Mr. Lara, a very respectable merchant of Tabasco. She at first was much alarmed, but after she had been with us a few minutes she became entirely reconciled.

The commandant of this place behaved to us in a gentlemanly way. Received our flag. Wrote a very complimentary letter to Capt. Thompson for his kindness and humanity to his countrymen that were prisoners to us. He ordered his soldiers to fill our water and also sent fruit and other little articles on board. From this point, we cruised along in the neighborhood of Vera Cruz but adverse winds retarded our progress much. On the 17th, fell in with and captured the Mexican schooner *Rafaelita* (Capt. Punentes) formerly the *Correo* that Thompson commanded on the Texian coast in the first part of the war. Saturday, the 15th, saw four strange sails. Gave chase; some went one way and some another. I gave chase to a brig and after a few hours came alongside of her, but unfortunately she was a French vessel in ballast. I then stood after our little prizes; at 9 p.m. came up with them and laid to for the *Invincible*.

At daylight, the small *Correo* was not to be seen, neither the *Invincible*. I then stood in for Tampico and remained twenty-four hours in the bay. Found one English schooner at anchor. On the afternoon of the 21st, whilst standing to the northward with the *Correo*, saw two sails to leeward. Immediately chased and on nearer approach found one was the *Invincible* in chase of the other. I carried all sail in pursuit of her and at sunset came up with her at the bar of Tampico. She, however, proved to be an American schooner from New York bound to Tampico. Next morning saw another lofty sail. Went in chase of her; came alongside of her at 10 a.m. but she was French and Capt. Thompson did not wish to take him. I put my Mexican prisoners on board of her and sent them to Tampico.

Not finding anything that would be profitable, we again made sail for Matamoros. Found two American vessels at anchor outside. They informed us that the Honorable William H. Wharton had escaped from bondage. The sea becoming very rough rendered it impracticable for us to land or I verily believe we should have taken the Brazos de Santiago. We being run very short of water made sail and at 6 p.m. on the 25th took our departure from that place for Texas.

Nothing material occurred on the passage. At 5 p.m. on the 26th, passed Velasco and at 9 anchored at Galveston bar in 5 fathom water. 27th of August at 7 a.m. saw two sails in the offing. I immediately got underway and stood out for them. Found one was the *Invincible* and the other an American brig bound in. Capt. Thompson ordered me to send my boat on board. I done so and she returned with the Honorable S. Rhoads Fisher. At the same time, he ordered me into port and to take the *Correo* with me. At 10, came to anchor off the navy yard, furled sail, squared yards &c &c.

At 12, two suspicious vessels came off the bar chasing the *Sam Houston* into port. At 1, Col. Thruston came on board and stated that the vessels were Mexicans. I immediately sent my boats to all the vessels in the harbor and obtained a number of volunteers and some water. Got underway and stood down the harbor but unfortunately the water being low, the vessel run aground. The steam boat *Branch T. Archer*, Capt. Ross, with his usual promptness, came alongside and endeavoured to extricate us from our disagreeable situation. In the act of doing so, the hauser unfortunately passed across the stern slightly touching the rudder; instantly carried it away as it proved very rotten.

I immediately ordered all the men armed to go board the steamer and endeavour to board one of the enemy's vessels who were then in action with our brave little *Invincible*. As soon as the Mexicans discovered us, they hauled their wind and stood off to sea. The *Invincible* unfortunately lost her rudder and got on shore inside the bar. We used our exertions to save her but all proved ineffectual as she could not stand any kind of thumping. She soon went under.

Too much praise cannot be given Capt. Ross for his prompt assistance on our trying occasion. I also return my warmest thanks to the Honorable S. Rhoads Fisher, to Cols. Thruston and Graham, Capt. Hoyt and many other gentlemen who kindly and promptly volunteered their services on board the *Brutus*, not forgetting my own officers and little crew.

After the loss of the *Invincible*, I returned on board my vessel and after considerable exertion got her once more afloat and towed up to the Navy Yard and moored in safety.

I would respectfully recommend to the Honorable Secretary of the Navy the following officers for their bravery on various occasions, and also for their general gentlemanly deportment during the cruise: Lieut. Francis B. Wright, James G. Hurd, Dr. Francis T. Chrisman, Purser Norman Hurd, Midshipmen E. P. Crosby, D. H. Crisp and Robert Foster. The boatswain and gunner have also behaved themselves to my satisfaction and with much credit to themselves. I fear I have to regret the loss of the gunner, Mr. J. Dearing, also my clerk, Mr. Reed, and four others who were in the little *Correo*.

> I have the honour to remain, sir,
> very respectfully your obedient servant,
>
> J.D. Boylan
> Comd. Schooner of War *Brutus*

Galveston, September 1, 1837

I find it a very difficult piece of work to get our rudder unshipped as it is woodlocked 6 feet below water.

(MS, Navy Papers at Texas State Library)

CRUISE REPORT: YUCATAN CAMPAIGN OF INVINCIBLE & BRUTUS

TO THE HONORABLE NAVAL DEPARTMENT

SCHOONER OF WAR *BRUTUS*
GALVESTON BAY, 29TH AUG 1837

I have the honor of making a brief report relative to my cruise in command of the Texian fleet.

I left this place after my return from convoying the schooner *Texas* down to Matagorda on the 11th June, after despatching a boat from the *Brutus* in charge of a warrant officer with letters and despatches from S. Rhoads Fisher Esq. The officer and boat having returned on board about one o'clock in the morning without any reply from shore, and my fleet lying to anchor in an open roadstead and much exposed to sea, I saw proper to get underway and stand to sea. I proceeded to cruise off the Balize where we remained two days without success. We then shaped our course for the Campeche Banks, and for fear of a separation of our small fleet, Capt. Boylan and myself agreed to make our rendezvous at the Island of Mujeres, where we met and filled up our water.

During this time we despatched our boats to the Island of Cancun where we supplied ourselves with turtles that benefited us much and saved our little stock of salt provisions. Nothing of value was to be obtained. We then bore away for the Island of Cozumel which we found to be one of the most desirable places in all the course of my travels. We immediately, after coming to anchor, went on shore and took possession under a salute of 23 guns, and with a hearty welcome by the inhabitants on the seaboard.

We surveyed the Island as well as circumstances would admit and still became more infatuated with its delightful situation and the salubrious trade wind which blows without cessation.

This connected with the beautiful roadstead and anchorage and the richest of soils which produces the finest kind of timber and that of a variety, induces me to think, not only think but am well convinced that it will become one of the greatest acquisitions to our beloved country that the admiral aloft could have bestowed on us. I hoisted the Star Spangled Banner at the height of forty-five feet with acclamations both from the inhabitants of the Island and our small patriotic band, the crews of our two vessels. We then filled our water and made sail on our homeward bound passage, passed the Island of Mujeres, falling in with and chased some of which we could not come up with on account of our dull sailing vessels—which I can with an open heart assure you is nor never was what they have been cracked up to be, or we would have taken many more valuable prizes. For be assured no exertions on the part of our officers [is] wanting. The course I pursued from thence westerly until we got off Telchac on the Yucatan coast, where we received a rascally reception from a squad of cavalry who rushed from behind the sandhills with their escopetas, pistols and cutlasses, and fired on Judge Fisher and Capt. Boylan of the *Brutus* who went on shore unarmed with the exception of a pair of pistols which Judge Fisher had taken with him, and which the Judge nobly discharged amongst their cavalry and fortunately hit one of the riders, who fell forward on his horse's neck and from that to the ground.

Previous to this we had acted with every degree of humanity towards our enemy, but from that time our feelings became excited and I gave an order to this effect—to burn, sink and destroy everything we can athwart and then commenced burning their towns, names at present I cannot remember.

We then shaped our course for Sisal, which we engaged under many disadvantages on our part—small guns, bad powder, wants of shells, port fires, Congreve rockets, small arms &c. We stood them a fight of two hours and forty minutes, close under their castle which was mounted with six pieces of cannon, calibres from eighteen to thirty-two pounders, independently of some guns they had planted behind the bushes which opened a severe fire upon us from a quarter we least expected. I, finding nothing was to be gained from the enemy and my vessel lying in fifteen feet water, thought it best to weigh anchor and get to sea to save our vessels and men.

We received no damage from their shot, but directing ours in as well as we knew how, be assured that few of ours was wasted. The old Commandant and his house smelt a little of the devil and gave up his ship, hoisted his jib and wore round. But the hot shot which we fired was merely cold shot warmed and did not have the desired effect or we could have burned the whole town which consisted in about one hundred houses. I from thence took my departure westward over the banks and close along the enemy's coast, and from thence to the Isle of Alacranes, where we captured the Mexican schooner *Abispa*, having the schooner *Telegraph* in company and sent them both to Matagorda.

Aug. 8th, while lying in the Alacranes, at daylight in the morning, saw a suspicious sail standing in for the land. I made signal to the *Brutus* to get underway and chase. In the afternoon, the *Brutus* returned bringing in the vessel, which proved to be the British schooner *Eliza Russell*, Capt. Russell from Liverpool bound to Sisal, and from documents found in the possession of the captain, found that nearly half her cargo was Mexican property shipped by a Mexican house in London and consigned on account and risk of J. G. Gutieress of Yucatan. From the tenor of these documents, I felt it my duty to order her to Texas to have this Mexican property condemned.

From the Alacranes I proceeded down along the Mexican coast, touching at Sisal, Campeche, Laguna and the intermediate ports of the coast. On the morning of the 12th August off Tabasco, fell in with and captured the Mexican packet schooner *Correo* from Vera Cruz from whom I learned that the Mexican squadron was at Vera Cruz. Anchored off Chiltapec and landed all our prisoners. The commandant on shore was polite and supplied us with some water and fruit. From thence shaped our course to Vera Cruz but adverse winds delayed our progress much.

On the 17th the *Brutus* captured the Mexican schooner *Correo*, formerly the war schooner commanded by Capt. Thompson, manned her and sent her away. Off Tampico chased several vessels, some we boarded, some we could not catch.

On the morning of the 25th, run close into the Rio Grande. Boarded two vessels but they were neutrals. From thence proceeded to Galveston. 26th, came off bar and laid by for the night. 27th at 8 a.m., I ordered the *Brutus* into port and take the *Correo* with her. It was done and I remained at anchor outside. Near noon, the masthead lookout reported three vessels standing in. Judging from their appearance that they were Mexican, I immediately got underway and beat up for them with all my guns and men prepared for close action, the cannon well charged with round, grape, canister and chain shot. I used every means to bring them to close engagement, when they tried to avoid it and get to the weather gage of me and to get me between the two brigs.

My only means of bringing them to action was to bring my little vessels in stays on the opposite tack—this brought them to range right ahead—I made all sail and closed with them as much as possible, but they at the same time kept their luff from me and I at length got within grape and canister distance and gave them one broadside from the larboard battery. This seemed to frighten them away. The leeward brig being a little astern, my only method was to helm and wear round. This relieved my starboard guns then under water, but equally well charged with all sorts of shot, and which we equally as well directed. The shot they received immediately put them to flight, for they hauled their wind that they might get close to each other, which they did, and then opened their fire on me, though to little effect.

Their shot flew over and under us in every direction, but not a man was hurt or a spar of ours cut away while our long tom spoke the Texian Language and almost every shot told well, and with a small assistance rendered to me the same two brigs-of-war would have been ours. But my having no aid could get no manly play from them. I, to save my vessel, hauled for the bar with a hope of enticing them into shoal water where I could have managed them better. They smelt the rat and as cowards sneaked off, we giving them the parting shot which helped to clear their quarter deck.

But what has distressed me much is that I should be so unfortunate as to lose my vessel on the bar although she never was the vessel she was cracked up to be.

We have now lying in this harbor the brig *Phoenix*, prize to the *Tom Toby* which we have examined and find her to be a well-built vessel that will bear a battery of fifteen guns. A pivot twenty-four pounder will be highly recommendable with fourteen side guns, medium eighteen pounders. The little schooner [*Correo*] will certainly be very serviceable to the Navy, her sailing fast and light of draft of water will enable us to reconnoitre the enemy's positions. She mounted with a pivot nine pounder will be sufficient for her. The *Brutus* and *Tom Toby* will make us a formidable navy which will be able to defend our coast and prevent the enemies intercepting our commerce.

The *Phoenix* and *Brutus* are both in want of rudders though there is a possibility of getting the *Brutus'* rudder scarped, although it is extremely difficult to get the iron-work done. Several of my crew have deserted and stolen one of my boats. Some of them for fear of being kept in the service have left under pretence of having nothing to eat, of which we have plenty for some time.

Should the government see proper to fit out the brig *Phoenix* as a man-of-war it will actually be necessary to examine her bottom as she has been ashore on the bar and thumped her rudder off, and no doubt has injured her copper if not her bottom. Be that as it may, the coast ought to have vessels that will carry their guns above water, which was not the case with the *Invincible*, for one side or the other in carrying sail was always under water and rendered the lee guns entirely useless. The schooner *Brutus* is leaky and wants sails and rigging and as well to be hove out.

The prize schooner *Correo* requires some sails and rigging. The particular dates of circumstances that occurred during our cruise is entirely out of my power to give, as all my papers have been washed to pieces by the sea breaking over the wreck, and I am equally barely stripped of clothing.

> Most respectfully
> Your obedient servant,
>
> H.L. Thompson
> Commanding, Texas Navy

(MS, Navy Papers at Texas State Library)

PRIZES OF THE
YUCATAN CAMPAIGN

To the Honorable Secretary of the Navy, Houston
On board the prize schooner Correo
Galveston, Sept. 20th 1837

Sir;

I have the honor to state in pursuance of your instructions, that I have examined and surveyed the brig *Phoenix,* schooners *Correo* and *Thom. Toby* and report as follows:

I find the brig *Phoenix* to be a new vessel; well built of Campeche cedar or iron wood, which is considered equally as durable as live oak; she is coppered and extra copper-fastened. Her frame cannot be surpassed for service and durability, the timbers being large and compact, more particularly the floor timbers between which there is no space. The transom is a double one and joined to the frame by knees extending forward over the four after futock timbers; her stern piece is very large and well supported; her kelson light, and it will be necessary to run a "back bone" through her whole length to add to her strength.

The decks are supported by large beams of lignum vitae, which rest upon stringers running the whole length from stern to stern (entirely dissimilar in this particular, from the usual construction, which is putting clamps under each beam, giving support to the beams alone) whereas the stringers give strength and support in this vessel to the whole frame, but especially to the floor and futock timbers.

As there is scarcely sufficiency of room between decks, I would recommend that the lower stringers be pared down even with the lining planking; the lower beams dropped two feet, and placed down upon the clamps, there never being a very great strain upon the lower deck. It will also

25

give the room which is essentially necessary between the spar and lower deck.

The upper deck frame is good; the planking is full three inches thick and fastened with a view to service and durability. The planks are of cedar. The outer planking on the bottom is three inches in thickness; that on the bends four, and on the side five inches. In recommending that the sleepers, which support the lower deck beams, should be removed, I would respectfully add that it can be accomplished at a trifling and insignificant expense, but by ship's carpenters alone, who possess every facility for work of this kind. It is also requisite that she be coppered two streaks higher, as she is now only coppered to light water mark; and as it is well known that the worms in the waters of Texas are more destructive than elsewhere, her bends would otherwise be in great danger from them.

As she has never been finished, she has no cabin or accommodation for officers, but room sufficient for the most ample apartments. The trunk now standing for her cabin should be removed, as more room, then, will be given at quarters, and less chance for splinters, which it is manifest, imperiously deserves the highest consideration.

Her upper frame is equally as strong as the lower. The staunchions are large, and mortices rest between the heads of the futock timbers. She will require new planking upon her bulwarks; that [which] she now has being too slight. Her spars are good; lower masts large, with long mastheads; her topmasts are slight and new; top gallant masts will be requisite, as those now on her are too short and out of all proportion. The yards although sufficiently square and full large for spreading the canvass upon her, can be replaced to very great advantage. Some of the sails are one-third worn and "middle stitched," the residue are new. It will be highly important that she should possess some new light sails as the canvass she can now spread is too scant for a cruiser.

The standing rigging is good, much better than new, as it has been stretched and has a sufficiency of tar among the hemp for its preservation. It is what is called the French-made-rope, which is not surpassed for durability by any other. I have had no opportunity of examining the eyes of this rigging, which would have required lifting to have done so. The running rigging however, is very much worn and will have to be replaced by new.

Her draught of water now, with 70 tons of ballast, is 6-1/4 feet aft and 5 forward. With ammunition, stores, provisions and water for a cruise of 4 months, with a crew of 150 men, which will be her complement, she will draw 9-1/2 feet aft and 7 forward, at all events not over ten feet. When taken she was laden to her supporters with salt, and drew 10-1/2 feet coming in. She fell out of the range of the channel (the Prize Master being unacquainted with it) and knocked off her rudder, which is about being replaced by one her owners have had built.

She has no boats, but one anchor and chain. She will have to be "hove out," her bottom examined and some repairs put upon her which ship carpenters alone can do. I herewith give her dimensions and tonnage.

Length from taffrail to night head, 96 feet
Length on deck, 93 feet
Breadth of beam at foremast, 21.10 feet
Breadth amidships, 24 feet
Breadth aft, 14.10 feet
Depth of hold, 12 feet
Height between decks, 4 feet

Tonnage by carpenter's measurement, 274 tons
Custom house measurement, 230 tons

She will bear a battery of ten, 12-pound gunnades, with two long twelves for bow chasers. From her narrowness above water it would be advisable to put on her a heavier battery as it would assuredly make her crank, to counteract which a larger quantity of ballast than would be necessary, would have to be put into her hold, the inevitable effect of which would be to cramp her and materially impede her sailing. Even in her present straight end conditions she could be fitted for a short cruise at trivial expense, but if at the outset the government, without regard to necessary expenses, equip her as an efficient man-of-war she will hereafter be at much less expense than if a temporary outfit be thrown upon her, requiring, continually, additional repairs and never in an efficient state.

From the model of the *Phoenix* she necessarily must be (if in proper trim) a superior sailer, and I learn from those in whom I can place reliance, and who have known her in her short career, that she has proved to be such.

In conclusion I would, most respectfully, assert that a vessel so well adapted for our purpose could not be obtained at this time. She is built of the best materials with the view to service and durability; an exceedingly fast sailer; large and commodious; light draught of water, and she cannot be considered dear at double the price her owners demand for her.

No. 2—The *Correo* is an admirable vessel of her class. She is sixty-five (65) tons burthen, Baltimore built and 2-1/2 years old, coppered and copper-fastened, built of Live Oak. Proved a fast sailer since she has been in our possession by sailing in competition with the *Brutus* and *Invincible*, and is well adapted for a *guarda casta*, or cutter. Will bear a battery of one long 18 as a pivot gun, and two 6s or two 9s as she now has mounted as side guns. Her present accommodations are ample for the officers she will require. A berth deck can be constructed for the accommodation of 40 men leaving sufficient room for provisions and stores.

From information and observation, I am induced to believe that her copper is somewhat broken but nothing however of material disadvantage. She will require a new suit of sails which should contain more canvass in them than those now bent, which are worn out. Her standing rigging is good; will, however, require overhauling. I would recommend that the top mast yards and sails saved from the wreck of the *Invincible* should be put upon her. She can be fitted out at a trifling expense and be made to last a length of time. Her dimensions are as follows:

Length from taffrail to night head, 67 feet
Length on deck, 62 feet
Breadth of beam, 19 feet
Depth of hole, 7 feet
Tonnage by measurement, 65 tons

I would recommend the long 18-pounder at Victoria as a battery which was formerly mounted on the *Viper* and bears the familiar cognomen of "Widow Porter," with the two brass sixes on board the *Tom Toby* which is the property of the government, or the two 9-pounder carronades now mounted on board of her. All the necessary alterations required can be effected where she is in a short space of time and at a very moderate expense.

No. 3—Schooner *Tom Toby*. I find her to be an excellent vessel perfectly seaworthy and copper-fastened, butt bolted; infinitely superior for service to the *Invincible* being built more firmly having much more room in proportion to her size. Her owners have spared no expense in her outfit being well supplied with rigging, sails and armament and all in excellent condition. Her battery at present consists of one long nine as a pivot gun, 4 nine-pound gunnades and 4 six-pounders as side guns. This battery is by far too heavy for her making her crank and it hinders her sailing.

She is now in proper trim for sea with the exception of provisions. She is seventy (70) tons burthen and can store provisions and water for a complement of fifty (50) men for a cruise of four months. In the present state of our affairs I would earnestly recommend her to be purchased as a vessel worthy to be taken into service. The only change she will require will be two streaks of copper as she is only coppered to light water mark. Her rigging is all seaworthy and in excellent keeping. Her sail is only one-third worn and have never been "middle stitched"; has two anchors and cables and one good boat. Her dimensions are as follows:

Length from taffrail to night head, 80 feet

Length on deck, 76 feet

Breadth of beam, 21.6 feet

Depth of hole, 8 feet

Height between decks, 4.4 feet

Tonnage by measurement, 90 tons

Custom house measurement, 70 tons

Drawing when ready for a cruise 8 feet aft, 6 feet forward.

As tis absolutely essential and requisite that an addition be made to our Navy without delay, I would earnestly and respectfully recommend the immediate purchase of these vessels. At any other time I should be loath to advocate the purchasing and equipping of vessels for the service which are not built expressly as vessels of war.

There are many reasons to be advanced to prove that such a measure would be detrimental to the interest of the government and service. These reasons are so apparent that I will not take the liberty of advancing them.

I have the honor to be
very respectfully
your most obedient
& humble servant,

FRANCIS B. WRIGHT
Texas Navy

(MS, Navy Papers at Texas State Library)

Voices of the Texas Navy

THE CASE OF THE BRIG POCKET
BY C. T. NEU
originally published in the TSHA *Quarterly,* April, 1909

THE CAPTURE OF THE *POCKET*

In March, 1836, when Texas was engaged in a life and death struggle with Mexico, and when the Texans were particularly anxious to gain the good will of the government and the people of the United States, an event occurred which might have resulted in alienating the sympathy of that nation had the Texan authorities not taken immediate steps to correct matters. This was the capture of the brig *Pocket,* a vessel sailing under American colors.

An account of the capture was given by Aleée La Branche, the United States *chargé d'affaires* to Texas, in a letter which he wrote to R. A. Irion, the Secretary of State of the Republic of Texas, on November 29, 1837. He says:

On March 20th of 1836, the brig *Pocket,* sailing under American colors and belonging to citizens of the United States, left New Orleans for Matamoros. On the voyage, she was captured by the Texan armed schooner *Invincible,* commanded by Jeremiah Brown, and carried to Galveston, and her cargo appropriated without trial or condemnation by persons acting under the authority of the Texan government. The captain and the crew, with the exception of the second mate, who was still more severely dealt with, were detained nineteen days at that place, after which they were released and suffered to embark for New Orleans. Permission was given them to take such articles of private property as belonged to them, but after a general search they were unable to find anything. Their clothing, hats, books, quadrants and charts were all missing, having been

33

already secured by the captors. Previous to this the passengers were transferred on board a Texan armed schooner called *Brutus*, where they were stripped and searched by a person named Damon, who acted as lieutenant, and four of them, viz., Hill, Hogan, Murje and Campo, were immediately put in double irons by him. One of the passengers, Taylor, had his trunk broken open by this Damon and four hundred ninety-seven dollars ($497) together with other property taken therefrom, amounting in value in all to eight hundred dollars ($800). When he desired to obtain a simple receipt for the money alone, he was put in double irons.

Hogan and Campo received one hundred lashes with a cat-o-nine tails, stretched on an eighteen-pound cannon and were threatened by Hurd, acting as captain of the *Brutus*, and Damon, that they should be hanged; the foreyard was accordingly loosed and braced for that purpose, and the offending victims were actually brought on deck with ropes around their necks and tortured with their impending fate. Somers and Taylor were kept in irons, the former for the space of twenty-five days, and the latter for seven weeks. At the expiration of these periods, instead of being released, they were forcibly detained, without any legal pretext or excuse, for upwards of four months and seven months separately, when they were permitted to depart for the United States. Somers during all this period was compelled to perform various work, such as unloading vessels, etc., and had all his clothing and instruments of navigation taken from him.

In the same letter, La Branche also gives an account of the seizure of another American vessel, the *Durango*, which at about the same time as the capture of the *Pocket*, was seized at Matagorda and pressed into the service of Texas by the orders of John A. Wharton, adjutant general of the Texas army, and William S. Brown, commander of the Texan armed schooner *Liberty*. The claims for both vessels were usually urged together, and when matters were finally settled, provision was made for the payment of an indemnity for both together.

La Branche's account gives only one side of the affair; it is also somewhat prejudiced. The treatment of the crew and passengers was not at all as brutal as he made it appear. In fact, Captain Howes of the *Pocket*, the first officer, and several of the crew made an affidavit in New Orleans to the effect that while they were under the control of Captain Brown, they were treated with kindness and respect. Alexander Humphrey, a passenger on the *Pocket*, made a statement to the same effect to William Bryan, the Texan agent at New Orleans. He also stated that no part of the cargo went to the crew of the *Invincible*.

The true facts in the case seem to have been somewhat as follows: Captain Brown, in the exercise of the belligerent rights of Texas, was cruising against her enemies and attempting to enforce the blockade of Mexican ports. The *Pocket* was bound for Matamoros, a Mexican port, and when she fell in with the *Invincible*, her captain refused to show his papers. Captain Brown then boarded the brig, compelled the officers to deliver up the papers and examined the cargo. The examination disclosed the fact that the *Pocket* was sailing under false papers and that the cargo did not correspond with the manifest and papers showing her clearance from the custom house at New Orleans. There seems to be no doubt that the cargo consisted of contraband of war, this fact being clearly brought out on the trial of the crew of the *Invincible*. There is some conflict of statement as to the articles composing the cargo. It is certain that the *Pocket* was carrying provisions that were intended for the Mexican army, and Captain Brown stated that powder, ammunition and other military stores were found on the brig.

But this was not all. A further examination of the papers revealed dispatches to Santa Anna, containing information that would aid him in his operations against Texas. He was informed of the force on each of the Texan vessels and instructed as to the best mode of attacking the Texans on land. There was also included "a chart of the whole coast, minutely and beautifully laid down—all surroundings, etc."

On board the *Pocket* were also several persons who were in the Mexican service, among them the notorious Thompson, who had only a short time before been imprisoned at New Orleans on the charge of piracy. This was the same Thompson who, while endeavoring to enforce the Mexican revenue laws, had been so insolent of the Texans at Anahuac. In September of 1835, he attacked the schooner *San Felipe*, a vessel owned by citizens of the United States, but was himself captured by the *San Felipe* and carried to New Orleans to answer to the charge of piracy. With him at the time of the capture was a lieutenant of the Mexican army, Don Carlos Ocampo. They were released on January 15, 1836, but Thompson was immediately rearrested by his creditors. But their affairs were apparently soon straightened out and both were returning to Mexico on the *Pocket* when it was captured. With them were Hogan and Taylor, officers of the Mexican navy. This probably explains how the papers describing the Texas coast came to be found on the *Pocket*—Thompson may have collected the information contained in them while he was stationed at Anahuac.

The conduct of the Texans after the capture was set forth by Samuel Ellis in a communication to the editor of the *New Orleans Bee* in May of 1836. He said:

You assert that the cargo was American property and actually belonged to Lizardi & Co. until delivered. The evidence of one of the firm, given before the examining court, was that the cargo on shipment was by the order of and charge to Rubio & Co.; that the premium was charged to them and that they considered the cargo at their risk. That such was the understanding is evident from the clause of the charter which expressly stipulates that the brig shall carry a signal generally known as that of the acknowledged agents of Santa Anna, which signal was to be furnished by Lizardi & Co. As further proof of the character of the vessel and the purpose for which she was engaged, we have the evidence of three witnesses on the trial that Captain Howes acknowledged to them that he was engaged after his arrival at Matamoros to transport Mexican troops to Texas...

On the arrival of the *Pocket* at Galveston she was, by the evidence of the captain and crew, given over to the Texan authorities and the allegations in [Captain Howe's printed] protest, which carry upon their face the appearance of oppression, were made under the direction of, and by the order of the Texian government, and being out of the jurisdiction of the United States and perpetrated by a government *de facto*, that government alone is responsible. Almost every allegation made in the protest is proved to be false...by the proof given on the trial. Several witnesses deposed as to the extreme delicacy used in the examination of the baggage of the passengers, and that American property was in every instance respected. So far as regards the treatment of the crew while on Galveston Island, being put into a tent on the beach and being short of provisions, the president of Texas was at the same time living with his family under the same shelter and equally destitute. The refusal to admit him on board his own vessel was caused by his own conduct, of which ample evidence is given.

In regard to the money handed by Mr. Taylor to the Secretary of the Navy, and by him handed to the purser, the Secretary was not the person to [issue a] receipt for it. Mr. Taylor being impertinent and troublesome, was ordered forward in charge of a marine, but was not put in irons. The money was held subject to his order, and has been or will be, restored to him when demanded.

The second mate Somers was one of the passengers put on board at New Orleans. He held a commission as lieutenant in the Mexican navy, and was furnished with funds by the Mexican Consul, as was proved by evidence on the trial. His name was not on the roll of the crew and he was well known as an enemy and a spy. The other passengers, excepting those well known to be Mexican officers, were treated with attention and respect, and the amount of their passage in the *Pocket* and in the *Congress* to New Orleans, together with all damage sustained by them has been paid by the government.

The capture of the Pocket, whatever the results thereof, was a very fortunate event for the Texan army. The cargo, consisting mainly of provisions, was, according to the 1860 Texas Alamanac, "a most timely assistance to the victors of the field of San Jacinto, who, short of provisions for themselves, were thereby enabled to retain the prisoners taken at that decisive victory."

Effects of the Capture on the Americans

When the news of the capture reached the United States, it caused much excitement, especially at New Orleans. Some looked upon the act as one of piracy. William H. Wharton, who was the in the United States as a member of the first commission sent out by Texas, was very much wrought up over the matter. On April 9, 1836, he wrote to the government of Texas, saying: "There is some talk of piracy having been committed by one of our vessels. In the name of God, let the act be disclaimed and the offenders promptly punished if such be the fact. I called on the Secretary of State this morning. He had not heard it officially." The charge of piracy, however, was soon discredited, but the affair brought home to the Americans the insecurity of their commerce on the gulf. The *New Orleans Bee* voiced the sentiments of those merchants who were not so much concerned over the struggle between Texas and Mexico as they were over the security of their commodities. A few quotations from the *Bee* [in May of 1836] will show how they viewed the matter:

It is high time that American commerce in the Gulf of Mexico should be protected from both Texas and Mexico, and unless the government interpose, the evils will be very serious...Our commerce should be protected from all.

The lesson...should not be lost on our Texas friends. It is neither the duty nor the interest of Texas to interfere with Mexican commerce...As much as we love Texas, we love America more, and cannot connive at any violation of American rights and commerce by Texas...

We have been shown a declaration signed by two captains of Texas vessels, Brown of the *Invincible* and Hurd of the *Brutus*, that they do not purpose hereafter to attack an American vessel or any ship belonging to American citizens. This was necessary to calm the apprehensions of the public, as insurance companies and merchants of extensive trade with Mexico were at first firmly resolved to send to Europe for goods ordered from Mexico and have them shipped to Mexico in French and English bottoms as the American flag was no longer respected.

Whether the action of Texas is or is not piracy, they should forego it in order to secure the energies of their friends and prevent the efforts of their enemies...

We are in favor of Texas liberty but not in favor of Texas capturing American vessels...

Of what use is the Star Spangled Banner if it cannot protect us from the depredations of a petty state creeping into existence?

William Bryan, the general agent of Texas at New Orleans, in a letter which he wrote to the president of Texas on May 14, 1836, shows the gravity of the situation. He says: "The result of the whole trouble will satisfy you as to the policy of invading the American flag. It would require but a few such instances as that of the *Pocket* to turn the government of the United States against you and stop every expedition in favor of Texas."

Case of the Brig *Pocket*

Court Proceedings in the Case

1. *In Texas*—The first question that arose in Texas related to the disposition that should be made of the prize. The exigency required the action of a tribunal of admiralty jurisdiction. As Texas had declared her independence only a month before, the government was still in some confusion, and the machinery of justice had not yet been put in working order. Robert Triplett, in his letter to Burnet of April 9, had recommended a decree establishing an admiralty court. But the government had ere then acted by establishing at Brazoria a district court with admiralty jurisdiction. On April 12, Burnet wrote to Collinsworth:

> A prize has been brought to Galveston by Captain Brown. The government has passed a decree to establish the district court...We want an able judge in the district where the trial must take place. Will you then accept the office of district judge for the district of Brazoria?

But it seems that Collinsworth did not accept the position, for on June 15, we find Burnet writing to Judge Franklin as follows:

> The ordinance establishing the district court for the district of Brazoria and your appointment under that ordinance were measures produced by the present exigency of the country which requires the action of a tribunal of admiralty jurisdiction. The capture of the *Pocket* produced that exigency, and the principal object of the early organization of your court was that the questions arising from the capture might be promptly and equitably determined, for it was known that the capture would produce great excitement in the United States. Several weeks have elapsed and no proceedings

have as yet been had on that important subject. The character of Texas and her interests are daily suffering and the evils admit of no relief but by a just adjudication at your bar.

Thus there was much delay in having the trial, this letter being written almost three months after the capture of the *Pocket*; but Judge Franklin was not responsible for this, for on June 4, William H. Jack wrote to J. K. West:

Owing to unavoidable accidents, it has been impossible to have a trial for the prize *Pocket*. It is likely to be determined soon.

Just when the adjudication took place is not known, but it was probably some time in the latter part of June or the early part of July. It is known, however, that as a result of the trial, the *Pocket* was adjudged a lawful prize. On October 27, 1837, R. A. Irion, the Secretary of State of Texas, wrote to William Bryan:

Shortly after my note to you relative to the prize brig *Pocket*, I saw Ex-president Burnet, who informed me that the adjudication took place before Judge Franklin, who had been appointed admiralty judge, and that the court condemned the brig as a lawful prize...There is no doubt of the decree having been made.

On what grounds it was condemned is not known, but from the character and cargo and papers found on the *Pocket*, the step was amply justifiable.

2. *In the United States*—In the meantime, the United States authorities had taken up the matter, for, as we have seen, the capture was considered an act of piracy. After the *Invincible* had brought her prize to Galveston, she proceeded to New Orleans, but owing to the excitement over

the capture of the *Pocket* she could not remain there with safety. On April 18, 1836, Bryan wrote to Burnet:

We have been compelled to order the *Invincible* back to Galveston; the capture of the brig *Pocket* is considered by the authorities as an act of piracy. The friends of Texas are among those in authority, and information was given me of the intention of the marshal to take the vessel and arrest the crew. We acted instantly and sent down a supply of provisions and ordered the vessel back to Galveston. We presume she has escaped. Captain Brown is out of the city and will probably not be able to join his vessel. Should she be detained, the cause of Texas will have received the severest blow she has yet met and the agency will be involved in trouble it will be hard to evade. Our situation with all the wealth and power of New Orleans arrayed against us is one of peril and danger.

But the *Invincible* did not get away. Commodore Dallas of the United States Navy, at the request of the insurers of the cargo of the *Pocket*, sent out the sloop-of-war *Warren* to seize her. This was done on May 1, and the crew of the *Invincible* was lodged in jail by the United States marshal, and held to answer to the charge of piracy. Bryan at once employed the ablest counsel he could secure and had the Texans brought to an examining trial as soon as possible. An examination was held on May 5, but for want of evidence the trial was postponed several days.

In the meantime, the seamen were confined in a prison which a Texan sympathizer characterized as "a dungeon, the exact model of the 'Black Hole' "

The trial lasted three days, and on the night of the third day, the crew was liberated. Justice Rawle, who tried the case, did not think there was sufficient evidence to justify a trial by jury. It appeared that no criminal act had been committed by the prisoners, as it was shown that the *Pocket* contained contraband articles that were intended for the Mexican army in Texas under Santa Anna. No act of malignant hostility had been committed and, of course, no piracy. After their release, the crew was cheered at every step and had supper given them and free admission to the theater. This kind of treatment showed that the mass of people in New Orleans was not turned against the Texans by reason of this unfortunate occurrence. There was an attempt on the part of the prosecutors to have the *Invincible* again seized and taken to Key West for a new trial, but nothing came of this.

The Texan sympathizers alleged that the imprisonment of the crew was brought about through the influence of Santa Anna's friends in New Orleans. They said that Lizardi & Co., who shipped the cargo, were the known sub-agents of Santa Anna; they were strengthened in this belief by the fact that the cargo was consigned to one Rubio, who was said to be Santa Anna's general agent and banker. It was also alleged that the Louisiana State Marine and Fire Insurance Company, which had insured the cargo for Lizardi & Co., was attempting to aid the Mexican cause, for it was at their request that the crew of the *Invincible* was seized. It was pointed out that the insurers would not need to pay a cent of insurance if they could prove that the cargo was contraband of war. But the company did not attempt to establish that fact. Instead, they sent to Pensacola to get a United States warship to seize the *Invincible* and this, too, when they knew that a civil officer could just as easily have taken charge of the vessel. The inference, therefore, was that they did not wish to have the fact established that the cargo was contraband, and were attempting to aid the Mexican cause.

But such was not the view taken by all. The *New Orleans Bee* for May 16 has the following to say on the matter:

The *Mobile Chronicle* says the *Invincible* was captured at the instance of Santa Anna's agent in this city. Santa Anna has no agent in this city; nor has the Mexican government any commercial agent here. Lizardi & Co. are not agents; they deal with merchants only. In the case of the *Pocket*, they received an order from a Mexican merchant. They were not bound to ascertain the purpose to which the goods were to be put. If there were any articles on board which did not appear in the manifest of the cargo shipped by them, to the captain, not them, belongs the responsibility. The goods belonged to Lizardi & Co. until delivered to the consignees. Hence it cannot be said that they were Mexican goods captured in a neutral bottom, but goods belonging to American citizens were captured in an American vessel. Why should Commodore Dallas be assailed for taking a vessel that captured an American vessel with American goods? Why should the insurance company be assailed for requesting that action on the part of Dallas in order to indemnify themselves and prevent future occurrences of a like nature? Why should Lizardi & Co. be assailed for sending goods on their own account to a Mexican merchant?

A few days after the liberation of the crew of the *Invincible*, the officers and crew of the *Pocket* arrived in New Orleans. Feeling was again stirred up and the Texans would have been arrested a second time and brought to trial but for the action of the Texan agents, William Bryan and Thomas Toby & Brother, who bought the *Pocket* and paid the damages sustained by the officers and crew. On May 14, Bryan wrote to the president of Texas:

With the assistance of our friends, Thomas Toby & Brother, we purchased the brig and paid her charter and demurrage. We have also been obliged to pay all the damage sustained by the officers and crew, amounting to eight thousand dollars ($8,000). This measure was absolutely necessary to save the vessel from the charge of piracy and maintain the public feeling toward the cause...The *Pocket* now stands as the property of Thomas Toby and Brother.

The purchase of the *Pocket* took place on May 10. Elijah Howes, the master of the vessel, on that day, in consideration of the sum of thirty-five thousand dollars paid him by the Tobys, Bryan and Hall, agents for Texas, executed a bill of sale of the *Pocket* to T. Toby and Brother. He was also paid fifteen hundred dollars for damages on account of the detention of the *Pocket* and gave a receipt for that amount on May 10. On the next day, he also gave a receipt for one hundred dollars for various articles of personal property taken from him by the officers and men left in charge of the *Pocket* by the government of Texas. Alexander Humphrey, John W. Waterhouse, C. Anderson and James Doherty were also paid for the damages sustained by them and the owners of the *Congress* were reimbursed for carrying the crew and passengers from Galveston to New Orleans. Thus it seemed that through the exertions of the Texan agents, matters were being smoothed out.

But trouble arose in another quarter. On May 19, the Louisiana State Marine and Fire Insurance Company instituted a suit in the United States District Court against Captain Brown of the *Invincible*, seeking to recover the amount of the premium they had been forced to pay Lizardi & Co. In the libel which they filed with the court, they set forth that they had insured the cargo of the *Pocket* for eight thousand dollars; that the vessel was bound for a port in Mexico with which republic the United States was at peace, and that in consequence of the unlawful capture of the *Pocket*, they had been forced by virtue of a policy issued to Lizardi & Co. to pay that firm eight thousand dollars. They then alleged that because of this payment, all right of action against the persons who had unlawfully seized the cargo was transferred to them, wherefore they prayed that Captain Brown should be forced to pay them eight thousand dollars. On the basis of this libel, Judge Harper of the federal court ordered that Captain Brown be held to bail in the sum of nine thousand dollars. On May 20, the United States marshal took Brown into custody, from whence he was released on the same day, having given bail with Thomas Toby and William Bryan as sureties. The case was to come up on the second Monday in December, 1836, but the record of the court does not show anything was done on that day.

The issue in the trial depended mainly on the legal condemnation of the *Pocket* by the Texas court. As early as May 16, J. K. West, the president of the insurance company, had written asking President Burnet to forward him copies of the condemnation of the cargo of the *Pocket*. Bryan also bestirred himself to secure the needed evidence in the case. On May 21, the day after he bailed Brown, he wrote to the president:

> It will now become the duty of the government to have the cargo condemned by a regular court, to have sufficient evidence forwarded of the character of the cargo, the documents and papers found on board proved as having been taken from the vessel, and all information you may judge necessary to forward to prove the legality of the capture.

There was much delay in forwarding the needed evidence, and it is probable that for this reason the case was continued. However, sometime before February 25, 1837, a judgment by default was rendered against Captain Brown, for on that date, the court:

On motion of Randal Hunt, ordered that the judgment by default be set aside, and that he [Hunt] be allowed to file an answer on behalf of the defendant in the case.

The attorneys for Brown then filed a plea to jurisdiction, averring that all questions relative to the adjudication of prizes brought into ports of Texas belonged to the tribunals and legal establishments of that country and none other; that officers of Texas war vessels ought not to be arrested in ports of the United States to answer for any capture or seizure made on the high seas; that vessels of belligerent powers may seize neutral vessels, take them into ports of their (the captor's) country to answer for any breach of law of neutrals, and the vessels of war are not amenable for such acts before any tribunal of the neutral powers; that the insurance company had arrested Captain Brown, but had in no manner alleged that the capture of the *Pocket* was made within the territory of the United States; that at the time of the capture Texas was a free and independent state, and Captain Brown was commander of one of her public vessels. For these reasons they held that the United States court was without jurisdiction in the suit and, therefore, prayed that Captain Brown be dismissed with his costs. Anticipating an overruling of this plea, the attorneys also filed an answer to the libel of the insurance company. They showed that at the time of the capture, Texas had declared her independence and maintained a government, and was, therefore, entitled to exercise all belligerent rights of a free and independent nation; that, under the authority of that government, Captain Brown had the right to cruise against the enemies of Texas; that in his capacity as captain he made a legal prize of the *Pocket*, whose cargo was enemy's property and intended as supplies for the Mexican army; that the *Pocket* was conveying hostile dispatches to the enemy; that she was sailing under false colors; and that her actual cargo did not correspond with the papers showing her clearance from the New Orleans custom house—all of which was in violation of the belligerent rights of Texas. They, therefore, prayed that the libel of the insurance company be dismissed.

But evidently nothing was done at the spring session of the court; for on August 4, 1837, Bryan again wrote to the Secretary of State:

> In my letters to the executive under date of February 22, April 12 and 21, I urged upon him the necessity of forwarding certain documents to save the loss of about nine thousand dollars for which I am bound in the United States court, being the value of the brig *Pocket*. Part of the documents were promptly forwarded by S. R. Fisher, Secretary of the Navy; those proving the condemnation of the brig as a lawful prize, the approval of the act by the government, and the record of the court condemning her under the great seal of state have never come to hand. The trial will come up in early December, and if such papers are not produced, the amount is lost...Will you do me the favor of forwarding such papers as are required or such as can be obtained in relation to this matter as early as possible, or advise me that they cannot be obtained that I may have time to prepare to meet nine thousand dollars in cash by sacrifices made to meet claims incurred for Texas.

Irion, the Secretary of State, at once exerted himself to procure the documents. He wrote to William S. Scott, the clerk of the district court of Brazoria County, requesting him to forward, with the least possible delay, a certified copy of the proceedings of the court in the case. Scott immediately transmitted the proceedings of the court, but unfortunately the decree of condemnation was not among the documents he sent. Irion at once forwarded to Bryan the documents he had received, but the attorneys in the case finding that they were not sufficient, obtained an adjournment of the case until the first Monday in January 1838. The decree of condemnation of the *Pocket* could not be found among the papers at Brazoria, so the only remedy was to enter the decree anew in open court at the next session of the district court. It seems that this matter was put into the hands of F. A. Sawyer, the attorney who had argued the case before Judge Franklin in 1836. On December 21, 1837, he wrote to the Secretary of State from Brazoria.

> As soon as I arrived here, I made out a decree on the back of the original petition in the case of the brig *Pocket* and sent it by express together with a certified copy to Judge Robinson both of which he signed, as was required, and returned to me. The original decree I have filed in the office of the clerk of the district court; the certified copy, dated about the 10th of December and signed by Scott, who was at that time clerk of the district court, and certified by Judge Robinson. I enclose to you and hope you have received it in time to go by the present trip of the *Columbia*.

Irion transmitted these documents to Bryan on January 3, 1838, saying:

> With regard to the brig *Pocket* I have at last succeeded in procuring all documents required by the memorandum of Mr. [Randal] Hunt which are herein enclosed to you. It was impossible to obtain them earlier. The matter was brought up before Judge Robinson, having alike with all other judges admiralty jurisdiction and under the laws the privilege of sitting on admiralty cases whenever occasion requires.

The evidence must have arrived too late, for the case seems to have been again postponed; at any rate it was still pending on February 28, 1839. On that date Bryan wrote to the Secretary of State:

On May 30, 1836, I advised the executive that I had bailed Captain Brown of the *Invincible* on the suit of the Louisiana State Marine and Fire Insurance Company for nine thousand dollars. The suit is now pending in the United States District Court. Has any diplomatic arrangement been made with the United States in regard to the settlement of the claims of the insurers of the cargo of the *Pocket*, or has any claim been made upon the government of Texas for the value of her cargo?

Webb, the Secretary of State, replied on March 15, saying:

There is no evidence in this department that any arrangement was made by the governments of the United States and Texas to settle claims of insurers or pay for the cargo. In private conversation with Mr. La Branche I have found out that the claims of the insurers of the cargo (and for which suit has been instituted in New Orleans) was not included in the treaty.

The case was evidently settled outside of the court, or simply died on the docket, for there is no record in the minutes of the court of a decision ever having been made. The last reference to it in the minutes is dated May 18, 1840, and states that the case having been called was continued.

Settlement by Treaty
with the United States

While the insurers' claims were pending in the courts of the United States, some of the officers and passengers of the *Pocket* were also clamoring for redress. One September 1, 1836, T. Toby and Brother wrote to the president of Texas:

We have just been waited on by the United States district attorney relative to a Mr. Taylor, who was a passenger on the *Pocket* and had four hundred and ninety-six dollars ($496) in money taken from him.

By January 1837, the matter had also come up before the United States government. Wharton, the Texan minister at Washington, on January 6, wrote to Austin that Forsyth had exhibited to him a complaint of the mate and other officers of the *Pocket* which should be attended to at once. Henderon, the Secretary of State, informed him that he could assure the government in Washington that the government of Texas would at any time cheerfully hear all complaints and give all such as were entitled to it speedy justice as soon as their complaints could be properly laid before it.

On November 29, 1837, in the same letter in which he gave an account of the capture of the brigs *Pocket* and *Durango*, Mr. La Branche demanded payment, not only for the vessels, but also an indemnity for the property taken from individuals on board these vessels, and for other injuries they had sustained. Irion, in answering this, informed La Branche that the president had been considering the cases, and that as regarded the *Durango*, he (the president) would recommend to Congress the passage of an appropriation for the amount demanded for it; but as regarded the *Pocket*, the circumstances of her capture and subsequent purchase by Thomas Toby and Brother rendered it improper for him to recommend a second payment. He would, however, recommend that a payment be made to Taylor. It will be remembered that the Texas agents paid Captain Howes thirty-five thousand dollars for the *Pocket*, which was insured by a New York firm, Barclay and Livingston. When the brig was captured, its owners called upon that firm for payment. Now, it was a question of whether the insurance company was obliged to pay the premium. Should they pay and then call upon the Texas government to reimburse them, the latter, should it comply, would have paid for the vessel twice. The Attorney General of Texas maintained that the payment made to Captain Howes was conclusive of the rights of the owners. He argued that the acts of the master of a vessel were binding on the owners, and that third parties, arranging with him as the accredited agent of the owners, should not be held responsible for losses resulting through his bad faith or inattention. Whether or not the insurers paid the premium does not appear.

However, the president soon changed front completely, and before March 19, 1838, he had waived all objections to the payment of the claims and had decided to recommend to Congress the payment of them all. It is probable that a desire to avoid any unpleasant relations with the United States induced him to give up his position. The United States *chargé*, La Branche, and the Secretary of State of Texas, Irion, then met at Houston, and on April 11, 1838, concluded a convention whereby Texas agreed to pay the United States government eleven thousand seven hundred and fifty dollars, which was to be distributed among the claimants. This sum, with interest accrued thereon at the rate of six per cent, was to be paid to the properly authorized person one year after the exchange of the ratifications of the convention. On May 3, the Texas Congress, in secret session, consented to and advised the ratification of the convention; and on June 14 the United States Senate did likewise. There was some slight difficulty in effecting the exchange of the ratifications as it was shown by Catlett, who was in charge of the Texas legation at Washington, in his letter to Irion of June 22, 1838. He says:

I have received the convention of indemnity to American citizens for losses sustained by the capture of the brigs *Pocket* and *Durango*. The Secretary of State from the first showed a disposition to accept the treaty and have it ratified on the part of the United States. But he seemed much in doubt whether it could be received on account of its not having been ratified by the president under the great seal of Texas, which was indispensable to a formal exchange. Another difficulty was that I had no specific powers for such purpose. But owing to the smallness of the amount and the unimportance of the matter, he yesterday waived these objections and accepted the treaty as it stood.

Texas was young in diplomatic affairs, and Catlett was no doubt somewhat embarrassed by the difficulties with which he had to contend. The exchange took place on July 6, 1838. The following day Catlett wrote to Irion:

In exchange for the copy of the convention which he placed in my hands, being the same as was transmitted by you, Mr. Forsyth gave me a formally ratified copy on the part of the United States. It was handsomely bound and had the United States seal attached. A similar ceremonial was expected to have been observed on the part of Texas, but that being impracticable, under the circumstances, he had agreed to waive any difficulties on this score as I mentioned in my last dispatch.

In this connection it is interesting to note that this was one of the only two conventions that Texas ever concluded with the United States.

On May 10, 1839, President Van Buren authorized La Branche to receive the indemnity and to give the necessary acquittal. On July 6, 1839, one year after the exchange of the ratifications, James Webb, the Secretary of State of Texas, turned over to La Branche a draft on the Merchants Bank of New Orleans for twelve thousand four hundred and fifty-five dollars, that being the amount stipulated in the convention with the accrued interest. La Branche on the same day executed a receipt for that amount giving

full acquittances to the government of Texas for all claims against said government of the United States for the capture, seizure and detention of the brigs *Pocket* and *Durango* and for injuries suffered by American citizens on board the *Pocket*.

The acquittances were to take effect as soon as the draft should be paid. On July 18, Robert Coupland of the Merchants Bank wrote to Webb acknowledging the receipt of the draft and stating that it had been duly honored.

Illustration by Dr. Alexander Dienst under the heading "Texas Naval Flags." See page 62 for a second illustration from the same collection. (Dienst Collection, Briscoe Center for American History.)

SEAFARING SAM CUSHING

In 1857, S. W. Cushing published his seafaring memoirs, *Wild Oats Sowings; or the Autobiography of an Adventurer*. Fortunately for Texans, Sam's sense of adventure induced him to serve as a middy aboard the *Liberty* in January, 1836. The *Liberty* was our first national vessel and was out on the Gulf making captures before our independence had even been declared.

Cushing, in the full text, claims to have entered this world on September 1, 1818 in Boston. As a boy, he devoured books and ate up the stories of his older cousins who served in the U. S. Navy. He determined at an early age that a life of salt water and danger would suit him just fine, and made his first attempt to that end at age 13. Witnessing a flogging on his first day, however, caused him to flee the scene and pursue a different brand of adventure where being scourged was not an option. He attached himself to a fishing smack and enjoyed a life of trolling for mackerel for a season. He then served aboard a trading vessel cruising the ports of the east coast for a few months, before attempting school again in his fourteenth year. His next encounter would be upon a "real square-rigger," a ship plying between Boston and Charleston. After one trip on this vessel, he associated himself with the *Thamar*, a smuggling vessel under one Captain Marks, bound for Haiti. He made two voyages to Haiti. During the second, he was caught by customs officials and found himself jailed and nearly executed.

Being barred from returning to Haiti, Cushing loafed about New Orleans and visiting New York. After his money ran out, he returned to the Crescent City in the fall of 1835 and worked as a bartender for several months. In January of the following year, he was approached by Captain William Brown to serve aboard the *Liberty*. The text which follows is what Cushing had to say about his time in the Texian navy. It is a unique description of the service as it gives us a glimpse into the human side of life in the navy, one of camaraderie and rivalry, inebriety and loyalty, patriotism and mutiny. His recounting of events is relatively consistent with history as we know it, with the occasional exception of his dates being a bit off.

Cushing ended up at the Battle of San Jacinto and stayed in the army, garrisoned at Fort Travis on Galveston Island. On August 21, 1836, he and several other former naval men hired a lawyer to pursue claims for them against the Republic for pay and prize shares for the period they served aboard the *Liberty*. According to other claims he filed against the Republic of Texas, he served in the army from May 10, 1836 to November 4, 1837, as a private in Captain Oliver's Company B. The fort, of earthen construction, in part built by Mexican prisoners of San Jacinto, was ruined in the hurricane of October 5, 1837, and this may account for the end of Cushing's service on the Island. He received his bounty certificate for 800 acres of land on November 4, 1837 and transferred it two days later to John Beldin in exchange for $75.00. When the land board met in February of 1838, he received 1/3 of a league for arriving in Texas in January of 1836 as a single man. Cushing took passage from Galveston to New Orleans on the *Virgil* on February 15, 1838. On the ship's manifest for that date, he is listed as being 27 years of age and a soldier. The age listed is obviously not consistent with Cushing's stated birth year of 1818, but such inconsistencies were not uncommon.

After his experiences in the service of Texas, Mr. Cushing went on holiday to England and France for a couple of months. Tiring of the climate there, he returned to visit his family in Boston, then off again he went, this time to Brazil. Sailing from Brazil to Uruguay, he sought employment there. Not wishing to participate in the ongoing Civil

War, he instead took a job on land, intending to work in the cattle trade. En route to his new location, he was reprimanded for violating a law prohibiting the chasing or hunting of ostriches within nine leagues of Montevideo, but with the help of some locals, he escaped arrest. He was not long at liberty before he was arrested and jailed. He was pressed into the service of Uruguay and fought aboard armed vessels on two sides of that affray until 1845. After losing an arm and quite nearly a leg, he was captured by the British and transported to England until it could be proved that he was a U. S. citizen.

What became of S. W. Cushing after his adventures in Texas and South America? It's hard to tell. Tracking down any evidence of a marriage or of the place and time of his death has proven fruitless. Seafaring souls rarely settle inland and with the use of only one arm, our battle-worn sailor may not have been fit for service anymore. There is a Cushing in New Orleans in the 1860 census—a native of Massachusetts, roughly our Cushing's age, living with a woman 20 years his junior. He is employed as an agent in that city, which occupation would have suited Mr. Cushing just fine. He fits our mysterious sailor's profile very well. If this is the same gentleman, he lived to the ripe old age of 78, expiring in New Orleans in January of 1896. If this is not the same gentleman, so be it. We're left with his memoir and his contributions to Texas history, both in living it and writing it.

Single starred banner illustration by Dr. Alex-
ander Dienst under the heading "Texas Naval
Flags." (Dienst Collection, Briscoe Center for
American History.)

CUSHING'S SERVICE IN
THE FIRST TEXIAN NAVY

...it was the 8th of January, the anniversary of the Battle of New Orleans, which is considered the most important holiday in the calendar among the creoles of Louisiana. "You are a stranger here, are you not, young man?" asked a very red-faced person, who was seated at one of the tables, carelessly turning over the dominoes which lay thereon. I answered in the affirmative. "May I inquire where you are from?" I gave the desired information, at the same time taking a seat at the opposite side of the table. "You are an American, I suppose?" I said that I was. "This is a beautiful morning; have you any objection to a walk with me before breakfast? I have something further to say to you." Always in search of an adventure, I assured him I had none. "I'll be with you in a moment," said he, rising and leaving the room.

As he did so I had a good opportunity to examine my new acquaintance, who was of a tall, commanding appearance and, as I saw at a glance, was no stranger to salt water. Sitting in the shade of a large Venetian screen, I had had no previous opportunity of noticing his dress, which now attracted my attention. It was of blue broadcloth, the coat of the long naval frock cut, with heavy gilt buttons stamped with the eagle and stars of the United States Marine, and garnished with shoulder straps of narrow gold lace. These, with a rich buff vest, boots and naval officer's cap of blue cloth with gold band, completed the exterior of Captain William Brown of the Texan schooner-of-war *Liberty*, then fitting out at Algiers, opposite the city.

This fact was communicated to me in a whisper by my friend Mills, as the gentleman entered the room. "You are caught," Mills whispered, as he sprang over the bar in answer to the call of the officer who very politely invited me to join him in drinking a julep.

"Are we all ready now?" he asked.

"Yes sir."

"Well, we will start by the way of the Place d'Armes; the artillery must be unlimbered by this time;" and as he spoke, the thunder of the first discharge reverberating through the streets brought a copious shower of glass upon our heads from the window of a large drugstore we were passing. The streets and squares were filled with people, all in their holiday attire, and apparently overflowing with patriotic ardor at the remembrance of the valiant deeds the performance of which they were assembled to celebrate. All was bustle and gayety, and we passed the Place d'Armes, which was filled by the Legion, through dense clouds of smoke, while the deafening roar of the artillery and the loud bursts of music from the band, intermingled with the cheers of the populace, who thronged every avenue to the place, formed a scene of uproar which could scarcely have been exceeded on the day they were striving to commemorate.

"That is brave work and all for a very good purpose," said my companion as we cleared the crowd; "but it is a pity that so much good powder should be used up, when it could be turned to so much better account."

"How so?" I inquired. "Do you not think it right to celebrate the anniversary of those days upon which our most memorable victories were achieved?"

"Most certainly," returned he, "and perhaps you will consider me a little selfish when I explain the reason of my remark, which was induced by the necessities of the cause in which I am engaged, rather than from anything like a want of patriotism or national pride, of which I lay claim to as large a share as any other man."

Here the good captain was interrupted by the ringing chorus of two long rows of darkies who were seated along the sides of a large tobacco and segar store, enlivening their toil with a song in honor of the day. Quite a large crowd had collected upon the sidewalk to watch the actions of the sable crew, who were engaged in the process of making segars, in the various motions of which they managed to keep time to their music. A little woolly-head, seated in one corner, sang the verse, after which all hands joined in the chorus:

> "I s'pose you've read it in the prints,
> How Packenham attempted
> To make Old Hickory Jackson wince.
> But soon his schemes repented;
> For we with rifle ready cocked,
> Thought such occasion lucky,
> And soon around the General flocked
> The hunters of Kentucky.
> O Kentucky, the hunters of Kentucky.
> O Kentucky—"

Here the little leader of the choir would show his ivories and, rolling up his eyes, shout, "Massa Jackson bery smart man!" All the rest in unison responded, "White folks say so, white folks say so. Eyah! Eyah!" It was a warm sunny morning, and the doors and windows were all thrown open, and through these, at the termination of each verse, went a shower of picayunes (6 1/4-cent pieces, the smallest coin then current in New Orleans) from the amused crowd on the sidewalk.

After a hearty laugh, we passed on to the ferry. Having got on board the boat, I asked my companion to let me know something of the cause he had mentioned as being engaged in, and in what way he was connected with it. "It is for that purpose I wished you to accompany me this morning," replied he; "and not only that, but I have hopes that I may be enabled to prevail upon you to join our side.

I have an idea that you can be of great service to me, as well as to yourself; therefore if you will join me at once, I will place your name upon the muster roll as midshipman. The bounty for entering the service is, for three months, three hundred and twenty acres of land; for six months, six hundred and forty; to be located at pleasure upon any unoccupied lands in the Republic of Texas. In addition to that, all who enter the service will receive the same pay per month as those of a corresponding grade in the United States navy. Such a chance as this seldom occurs, and I would advise you, if you love adventure and wish to make your fortune, to take hold at once, for I need your assistance very much."

Laying his hand upon my shoulder, he added, "Shakespeare says 'There is a tide in the affairs of men, which, taken at its turn, will lead to fortune.' This chance is yours. We will take a walk down to the *Liberty*, and then return to breakfast, and by that time you can make up your mind."

We had now reached the opposite bank of the Mississippi, at the place called Algiers, and half an hour's walk brought us to the vessel, which lay at Slaughter House Point, where she was being transformed from a peaceful merchantman into a man-of-war. A large force of carpenters, riggers and sailmakers were busy making the necessary alterations in her equipment. The armament, consisting of one medium twelve-pounder for a pivot gun, two long-sixes and two twelve-pound carronades, were being got on board by a part of the crew, under the supervision of the first lieutenant, a bustling Yankee named Walker. After surveying the schooner (which, by the way, was a very fine one of about eighty tons burthen,) and satisfying my curiosity, I rejoined my companion, the captain, who was by this time disengaged, and we returned to the city.

"Well, my lad, what do you think of the *Liberty* and of my proposal?" said Captain Brown, as we rose from the breakfast table. I answered that I was much pleased with the vessel, and in regard to his proposal, I inquired what particular duty he should assign me, giving him to understand that I had but little inclination to live on board the *Liberty* so long as I could enjoy the hospitalities of mine host of the *Fountain*. This objection was answered by his assurance that I was not required to go on board until the day of sailing. My employment, until then, would be to drum up recruits to complete the ship's company.

Perfectly satisfied with this arrangement, my name was duly inserted upon the ship's register and, receiving thirty-six dollars as advance wages, I went to work with a will. Having a pretty extensive acquaintance among the sea-faring community, I managed, by the 12th of January, to induce some forty men to enter on board of the *Liberty*, many of whom were stout able-bodied seamen, while the rest were Tennessee and Kentucky flat-boatmen—long, wiry fellows who were on their first visit to the city, and having been enticed to gambling dens had, by the turn of a card or roulette wheel, lost the proceeds of months of toil and exposure. These men, thus rendered desperate, were ripe for any adventure which promised to retrieve their losses.

While I was thus busy, the fitting out of the schooner was going on with great rapidity, and on the 12th we were all ordered on board and the *Liberty* hauled at once into the river. At 5 p.m. we dropped down the river, off the Freeport stores, where we let go our anchors and cleared the deck. At seven o'clock we fired our long-tom as a signal for our convoy to drop into the stream, ready to join the tow. In an hour more, we were underway and tearing down the river under a full head of steam. There were three tows in sight, plainly indicated by the glow of as many blazing furnaces brightly gleaming in the pitchy darkness.

Sailing Under the
1824 Flag in 1836

The morning light revealed the nature of our convoy, which consisted of seven schooners loaded with supplies for the Texan government. At 9 a.m. we cast off from the steam tug and came to anchor at the Balize. Here our colors were first hoisted, under a salute of thirteen guns. The flag was that of the Mexican republic—red, white and green, with the addition of the figures of "1824" in the center, expressive of the determination of the Texans to sustain the constitution of the Federal Government, which the usurper Santa Anna had overthrown. At 4 p.m., everything being in readiness, the convoy, in obedience to our signal gun, weighed their anchors and stood out to sea. It was near sunset ere we made sail, and the convoy was almost hull down in the distance, the vessels fairly staggering under the press of canvas. But though they had so much the start of us, our brave little craft was in the center of the squadron ere six bells (11 p.m.) had struck, using in the meanwhile nothing but the mainsail and jib. We carried the breeze with us until the afternoon of the 14th, when we made the low sandy coast of Texas, midway between the harbor of Galveston and the mouth of the Brazos. We now discharged our convoy, the coast being clear; they keeping away for the mouth of the river and we hauling up for Galveston.

Two light vessels, which were just entering the port, now attracted our attention by the strangeness of their movements. Suspecting they were enemy smugglers, we gave chase. Their apparent anxiety to escape us confirmed our suspicions and every rag that was available was soon distended to the breeze, and the *Liberty* flew over the combing rollers which skirted the bar. Ere an hour had elapsed, we were at anchor, our sails furled and our boats out, manned and armed, pulling lustily for the strangers who had taken refuge behind Pelican Island, where we could discern their masts while their hulls were hidden by the intervening island.

In the meantime, all sorts of speculations were rife in regard to the strangers, some pronouncing them pirates, others smugglers, and a few asserted they were some of the Mexican *guarda costas*. We already began to have indistinct visions of prize money and plunder, and anxiously waited the news from the boats. It was near eight o'clock when they returned, and Mr. Mayo, the officer in charge of the boats, had communicated the result. The vessels were engaged in the transportation of slaves from Cuba to Galveston, but such had been their expedition that the negroes were all landed before our boats had reached them. It is doubtful whether, in the state of the country at that time, they would have been adjudged as prizes to us, if this had not been the case. This, it will be remembered, was one of the speculations of the notorious Monroe Edwards, then located somewhere near New Washington on San Jacinto Bay.

Galveston Island and Bay presented at that time (January, 1836) a very different aspect to that which now greets the eye of the beholder. Where now stands an opulent city, one solitary wooden house presented itself to the eye. This, as if in mockery of the loneliness of the place, was dignified with the title of custom-house. A little to the right of it might have been seen the remains of the brick foundations of the houses which had comprised a settlement got up under the auspices of Lafitte, a former follower of whom we had on board our vessel. This man, whose name was Conti, at times told marvelous stories of money buried upon the island, and many were the rambling searches which I made in hopes of recovering the golden treasure.

In gazing at the shore, the eye in vain sought any inhabit-
ant in human shape, and yet everything seemed to swarm
with busy, noisy life. The red deer came bounding to the
brink of the wave, and stopping upon the glistening sands,
would stamp their tiny feet and toss their heads at our
unwonted appearance. The bay was literally covered with
wild fowl, while the low sandy point of Pelican Island, now
devoted to picnic parties, swarmed with countless thousands of the bird from which the island has been named.
The water, likewise, as if to give its quota to make up the
picture, abounded in immense shoals of mullet (by the
way, a most delicious fish,) with the beautiful redfish and a
larger species called by the knowing ones *granticoi*. These,
with occasionally a breaching porpoise or the shark with
his sharp back fins cutting the translucent wave, formed
an ever varying scene, most interesting to the spectator.

We were destined to remain here until our vessel was
commissioned and we had received orders to cruise. As
matters were not as speedily adjusted at the seat of government as they might have been, the most of us were fearful
that our stay in this port might be longer than agreeable to
us, or in keeping with the offers which had been held out
to us in New Orleans. As I have said before, our men were,
in fact, a mere party of desperadoes, and ere we had been
in Galveston a week, the murmurs of discontent became
both loud and general. Every possible effort was resorted
to, to quiet this feeling. Parties were started off to procure wood, and others to fish. The guns of the vessel were
scaled and their range proved, to the great discomfort of
every flock of pelicans or any living animal which made its
appearance upon the shore. Finally, tory-hunting became
the order of the day, and a party was started off to arrest a
man by the name of Parr, then living some eight miles east
of Point Bolivar. In this they succeeded, and that individual, together with his wife and children, were transferred

to the *Liberty* for safe keeping. But even this failed, in the end, to produce the desired result. Our commander was no disciplinarian, and the men in general were calculated to verify the old saying, "Put a beggar on horseback and he'll ride to hell."

Things began to wear a squally aspect when, on the morning of the 24th of January, the order was given to weigh anchor. All was now bustle and life. The idea of change was enough to restore the equanimity of all parties. When the word was given that we were to cruise upon the Mexican coast, the vessel fairly rang with yells of delight, forming a queer contrast to the formal precision and silent dignity of most national vessels. Having got the schooner underway, the prisoners, Parr and his family, were landed and cautioned as to their future behavior. The boats having been secured, we glided swiftly from the harbor and once more danced merrily over the blue waters of the Gulf. Our course was laid southeast for the coast of Yucatan, which we wished to make somewhere about Campeche. Our vessel, as I have before said, was a crack sailer, and with a fair wind made rapid progress over the Gulf. The men were in high spirits and all were on alert to be the first to discover a sail.

On the 28th of January, the loud cry of "Sail ho!" from a dozen voices at once greeted our ears. The vessel was barely visible upon the horizon. On the instant, every available sail was set and our course altered, having the wind of the chase, to cut her off. In a short time the guns were cleared, loaded and double-shotted. Small arms were distributed among the men. The articles of war for the regulation of ships of war were produced, and that part which refers to vessels in time of battle was read with great solemnity, much to the edification of the uncouth band who needed no spur to do battle to the death where plunder was in the question. In one hour, we were sufficiently near to see that all our excitement was useless, and our visions of bloodshed and booty melted into thin air.

The chase proved to be the bark *Montezuma*, then running as a packet between New York and Vera Cruz. Having spoken with this vessel, we again hauled our wind and resumed our old course, much to the satisfaction of those on board of the stranger, who had evidently been much alarmed at our proximity. The day after this occurrence, we encountered another stranger—an unwelcome one to the most of us—in the shape of a tremendous norther, before which we scudded under bare poles until the morning of the 31st, when we found ourselves on the Sisal bank (a large shoal lying off the coast) and the gale having abated, we again made sail.

On the morning of the 1st of February, we sighted the coast of Yucatan, and running well in with the land, succeeded in capturing a couple of fishing boats, for the purpose of gaining information of what was going on in the country. The crews of these vessels were mostly of the native Indian race, the descendants of the proprietors of the renowned Uxmal, the stupendous ruins of which have recently come to light. They were in a half-naked state, and seemed to be of a gentle, submissive race. Their boats were similar to those known by the French name *pirogue*, or dug-out, being hollowed from a section of the trunk of a single tree—a frail bark for a European or Yankee, but balanced and propelled through and over the waves with great dexterity by those primitive people.

At first they appeared somewhat alarmed, but finding themselves kindly treated, this feeling wore off and they became more communicative and, as far as they possessed any knowledge, it was rendered without much urging. Of the affairs of the nation they knew nothing, evidently thinking little and caring less about the political aspect of the times. They appeared to have an indistinct knowledge that their country was engaged in war, but with whom or for what purpose they were totally ignorant. One of our officers inquired of them whether they knew who and what we were. One of their number readily answered in the affirmative by casting his eyes around our hang-dog looking ship's company, and slowly ejaculating, "*Si, señor; caballeros del mar.*" This might very easily be interpreted as "seafaring gentlemen" or "pirates." Whichever he meant, the answer produced a loud burst of laughter from all quarters, while our commander, who loved a joke even at his own expense, pronounced the fellow a prophet. Having got all the information these people could communicate, they were allowed to depart, receiving salt beef and pork in liberal quantities in exchange for their fish, a trade which appeared to be very satisfactory to them.

We were now twenty-five miles south of Sisal, and the wind being very fair for that port, we made sail and coasted along until, at 3 p.m., we were in plain sight of the town and harbor, at the distance of about eight miles. In this position we supposed we were safe from a too rigid scrutiny, while we could make such observations as would be advantageous to us. Our flag was that of the Mexican republic and we had supposed the distance would prevent the figures in its center from being made out from the town, and thus we might pass for one of their own national vessels. But in this we were mistaken, as events subsequently proved. Some thirty vessels were at anchor within the harbor, which was but a mere indentation of the land, totally unprotected from the severe northers which vent their fury upon the coast of the Mexican Gulf. These were, with but one exception, vessels belonging to foreign nations, as their colors indicated. The vessel excepted was a large schooner anchored near the battery, from the main peak of which floated the flag of our enemy. As this craft very much resembled a Mexican man-of-war called

71

the *Montezuma*, which under the command of a Captain Thompson had committed some depredations upon the coast of Texas, she was at once adjudged a lawful prize by a council held in the cabin of the *Liberty*—that is, if we could get her.

This decided upon, we again made sail upon the schooner, and beat up under cover of the headland forming the northern point of the harbor, where we again hove to, for the purpose of making the necessary arrangements for our enterprise. While occupied with this business, a strange vessel made its appearance from the harbor, to which we instantly gave chase, with a view to acquiring some further news respecting the state of affairs in the town, and especially in regard to the object of our intended attack. The chase, on perceiving our maneuvers, immediately put back and attempted to regain the port she had left, but it was too late. She was very heavily laden and moved sluggishly through the water; and finding herself without hope in that direction, she put about, and crowding all sail, endeavored to escape seaward. This her captain soon found could not be done, and therefore making a merit of necessity, he hove his vessel to and awaited our approach.

When within hailing distance, the captain of the chase was ordered to come on board with his papers. The command was instantly but reluctantly obeyed by the captain, who had doubtless concluded that we were buccaneers, judging from his extreme agitation as he stepped across our gangway. He shivered like an aspen leaf in the breeze, while the pallor of his countenance showed mortal fear as his eyes rapidly took in the scene upon our decks. The bulwarks of our little vessel being low, a row of small iron stanchions had been attached to the top rail at intervals, and through an eye at the head of each was rove a ridge-rope to which was affixed a narrow tarpaulin curtain, fastened at the bottom to the rail. This contrivance served to give the appearance of hammock nettings and also concealed what was going on upon our deck. Whenever we had exposed ourselves to observation from the town and harbor, we had shown but just men enough on board our vessel to work the ship, the rest either being below or lying upon the deck under cover of the waistcloths.

Our people, at the moment of the stranger's advent on board, were mostly lying or seated around the deck, many using their best endeavors with files and scythe stones, to impart a keen edge to sundry cutlasses, bayonets, pikes, etc. which were likely to be brought into use before long. All this the poor fellow eyed with such an air of terror and bewilderment that he excited the ridicule of most of the men, some of whom called out to lead him along to the chopping block; another asked if he had said his prayers, advising him to do so without delay, as his time was come. Walker, the first lieutenant, who met him at the gangway, administered some comfort to the poor man by assuring him he had nothing to fear, adding that our vessel was the United States schooner *Nimble*, which had been sent to cruise for a pirate that had captured a vessel off the harbor of Pensacola. This little fabrication allayed the fears of our guest; and having been introduced to the quarter deck where a goodly array of decanters were standing upon the trunk, he soon became quite composed, and proved a more welcome customer than we had even hoped for.

From him we learned that the vessel we were so very solicitous about was indeed a Mexican one, but not, as we had conjectured, a man-of-war. According to his account, which, by the way, proved to be correct, the schooner had arrived that morning from New Orleans, and that at about noon boatloads of armed men from the shore boarded the vessel and, after some disturbance, had got her underway and anchored in her present position within pistol-shot of the battery which defended the town. This, though somewhat mysterious to us, was so far important to assure everyone that we should have to meet with resistance from those on board, let the case be as it might.

The battery mounted twelve long 18-pounders, and from their nearness to the vessel we so much coveted, and the almost certainty of finding the enemy on alert, the whole affair began to look somewhat hazardous. The sun had already dipped in the horizon when we parted with our new acquaintance, who left in a much more agreeable frame of mind than he possessed when he came on board, particularly as he was preceded by a fine ham and a case of choice liquors presented to him by way of recompense for his detention. As he left, however, the captain enlightened him of his intention to have the vessel, about which so many inquiries had been made, under our colors before morning. Wishing us success, he pulled away and, passing to his own vessel, ran down and announced his determination to lay-to during the night and see the sport. But he probably thought better of it, as the next morning we saw nothing of him.

We had been lying all this time within plain sight of the town, at the distance of about six miles. Having parted with our guest, we made sail once more and stood off shore, a course which we kept until it was quite dark, when we put about and made for the harbor under shortened sail.

The night breeze had set in strong from the northwest and the sea began to gradually rise. It was a fine starlit night, although light fleecy clouds were continually flying overhead. The crew were now mustered upon the quarter deck and the captain, in a short address, gave them to understand the object he had in view; namely, the capture of the vessel in question. This was to be done by means of the boats, as the appearance of our vessel in the harbor would put the enemy upon their guard. The announcement was received with every expression of satisfaction by the ship's company, who volunteered to a man to undertake the enterprise.

The work of getting out the boat, which was upon the deck, now commenced. In this we were unfortunate, on account of the heavy sea, which forced the boat, as soon as launched, with such violence against the side of the schooner, as to stave it in so badly as to render it useless. This misfortune did not dampen the ardor of the men in the least, and the quarter-boat next received their attention. Here too, we were unfortunate, for in lowering this boat, one of the ringbolts drew, letting the bow of the boat into the sea, where it was crushed under the counter of the vessel. We were thus obliged to cut her adrift and trust solely to the little cockleshell affair which we carried at our stern-davits.

Our expedition, in the first instance, was to have been composed of thirty men and officers. But our mishaps rendered it necessary to hold a council for the purpose of deciding whether we should attempt the capture or abandon the idea. The most sanguine were disheartened and it was decided, at the earnest solicitation of Walker, that the affair should be carried out. The party was reduced from thirty to fourteen men, including Mayo the sailing-master, and myself, as a special favor, as I was a great favorite with the first lieutenant. Had it not been for my light weight, the attacking party would have numbered one less.

By the time this arrangement was made, it was six bells (11 p.m.). The schooner had been all this time drifting in toward the roadstead, where the vessels were dimly seen at anchor and the sky had become overcast with heavy clouds, which prevented our being observed from the shore. Under these favorable circumstances, after taking a stiff horn of grog, we tumbled into the boat and put off, on our search for adventure.

For the first few moments things looked rather dubious, as the boat, loaded beyond her capacity, shipped water as fast as two of our number could bale it out with buckets, so that we expected every moment to be swamped. Those of us in the stern seated ourselves upon the top of the stern and quarters of the boat, our backs presenting a barrier against which the waves broke, which would have otherwise inevitably swamped us. In this way passed some ten minutes in almost total darkness, when we reached the outermost of the vessels at anchor in the roadstead. We rounded-to under the stern of a large schooner for the purpose of holding a consultation as to what course to pursue. A fine boat belonging to this vessel was floating beside us, into which part of our men immediately jumped, thus relieving us from the fear of becoming shark bait. Wet and shivering with cold, there was little time for deliberation and it was at once resolved to take possession of the boat we had just so unceremoniously boarded, by fair means if possible; if not, by foul. With this understanding, I was directed by Walker to shin aboard the schooner by means of a rope, or painter, which fastened the boat to her stern, and request the loan of her for a couple of hours.

"And in what way must I present so reasonable a request?" I asked, as I sprang forward to execute my mission.

"I'll leave that to you," said Walker, laughing. "Trust your wits, my lad, and bear a hand."

In another second I was upon the schooner's deck, reading the words "Sabine, of New Orleans" upon her stern, during the rapid transit. It argued ill for the shipper's sagacity to be lying at a single anchor in an open roadstead, with half a gale of wind blowing dead on the shore, without even the precaution of anchor watch. Everything was still, save the sharp whistling of the wind among the rigging. The companion-doors were closed and the slide shut over, and for an instant I paused to decide upon the best course to pursue in the negotiation which was about to commence.

Pushing back the slide and inserting my head, I shouted at the top of my lungs, "Hallo, below there!" I had not long to wait for an answer. My head was hardly withdrawn from the opening before a red nightcap popped into sight, followed by a pair of shoulders which would have done honor to a grenadier of the great Frederick's guard. Rather an ugly customer, thought I, as I confronted the apparition, which, after a few grunts and yawns, turned its head around and scanned the appearance of the weather. Then, facing me once more, he remained silent. "I'm in something of a hurry," said I. "Are you the captain of this vessel?"

"I believe I am," returned he. "And pray who the devil are you? How came you aboard of my vessel? And lastly, what do you want?"

"Did you observe the strange vessel in the offing yesterday?" I inquired.

"I believe I did, but what has that to do with my question?"

"Much, sir. That vessel is the Texan armed schooner *Liberty*, and her object is the possession of the Mexican vessel anchored near the battery."

I then related the situation we were in, in regard to the want of boats, and ended in making the request that he would assist us with the loan of his boat, to which he gave a flat refusal. I attempted to argue the matter with him but all to no purpose. And he ended by ordering me to leave the vessel. I now told him that his boat was already in possession of our people, and that he would find it hard to regain possession if he remained obstinate. Stepping aft, he was soon convinced of the truth of what I told him, and the sight of the armed men appeared to have a conciliatory effect upon him, as he immediately inquired when we would return the boat. I assured him it should be left at anchor near his vessel where he could easily recover it in the morning; and in regard to any trouble with the Mexican authorities, I hinted that he could excuse himself by asserting that the boat was taken by force.

Having thus secured this part of our objective, I told him that as he had extended aid to the enemy, it was likewise his place to afford comfort, giving him a short account of our perilous passage into the harbor, and the wet and cold condition of our party. In this, I succeeded in enlisting his sympathy to the tune of half-gallon of most excellent old Monongahela, which disappeared over the stern into the boat.

Everything being now ready, Walker, with seven of the party, took charge of our new acquisition, while Mayo and the remaining five kept possession of our own boat. The nearest merchantman to our object of attack was a large Hamburg bark, which lay about a quarter of a mile from both the Mexican and ourselves. It was decided that Walker's boat should gain the cover of this vessel and there await a signal from Mayo, whose boat was to make a circuit to the northwest to a point about equidistant with us, when the former was to dash at once for the starboard quarter of the Mexican, while the latter should endeavor to board upon the larboard bow. The signal was to be the flashing of a pistol.

All being arranged, we parted company and in a short time our boat gained the shelter of the aforementioned bark, where we impatiently awaited the signal from our comrades. Here for the first time, the Spanish watchword greeted my ear. The loud "*Sentinella alerta*" mellowed by the distance, sounded musically upon the water, answered by the more sonorous "*Alerta esta,*" the responding cry from post to post, delivered in various intonations of voice. It awakened all our energies for the desperate struggle which, from the appearance of things, we had a right to expect.

Our men sat in moody silence, their eyes strained in the direction whence we expected the signal, each moment, in our impatience, seeming an hour. The clouds now began to break away and scatter in drifts, revealing distant object more distinctly. The monotony of this scene was abruptly broken by Walker saying, "The devil! Here comes the moon!" As I turned, a small segment of the pale goddess of the night, partially obscured by light drift, appeared above the horizon. In a few minutes more, it again disappeared, much to our relief, and shortly after, the long wished-for signal was made by Mayo. It was instantly answered and, everything being in readiness, we let go and pulled cautiously with muffled oars toward the schooner.

We had arrived within one hundred yards before we were discovered, but now the loud *"Quien viva!"* of the Mexican sentinel fell harshly upon the ear, followed, before we had time to answer, by a volley of musketry. "Give way, boys! Nobody's hurt; give way!" shouted Walker. Amidst a shower of balls, the boat dashed alongside of the vessel, and in an instant, Walker, cutlass in hand, was upon her deck, slashing right and left and cheering on the party who had, in less than minute, joined our leader. How I got upon the vessel's deck I know not, but the first salutation I received was a blow from a stick of wood from one of the greasers, who had formed a line across the deck and were attempting to charge upon us.

At this moment the other boat struck the larboard side of the vessel, just abaft of the forechains. Mayo, with loud shouts, called upon his men to follow him. This was very opportune. A large part of the enemy turned to repulse the new assailants, while a few, pressed by us, jumped overboard and swam for the battery. Mayo's boat had lost its hold of the schooner. The greasers, being freed from any further alarm from that quarter, now turned in a body upon us, and for a few moments bravely contested the possession of the vessel; our men using nothing but their cutlasses, and the enemy their muskets, already unloaded their first attack upon our boat.

But such opponents as these were not long destined to stand against the athletic sons of the Mississippi Valley, who wrested a numbers of muskets from the hands of the enemy and soon put the whole posse to flight. A large number of them jumped overboard and swam for the shore, leaving us in possession of the vessel, and some dozen and a half of their number, part of whom were disabled or killed. "She is ours!" shouted Walker. "Bear a hand, lads, and loose the mainsail and jib." At this instant Mayo's boat, which had been swept far astern during the fray before the men could get their oars out, came alongside, and though late for the more desperate part of the enterprise, their assistance came most opportune.

One moment's examination told us we would have to depend upon our own wits to get the prize to sea. The mainsail had been unbent and taken on shore, while the main-boom was immovably fixed by means of a large iron bolt, which had been driven through its jaws and the mainmast. The headsails had followed the mainsail on shore, but the foresail—a large lug—yet hung in its brails. This was immediately shaken out, and the men were sent aloft to clear away the topsail and topgallant, who soon succeeded in loosing them, notwithstanding every footrope was cut.

While these operations were progressing, a few of us had managed to construct a temporary jib out of a large awning, which we found aft upon the quarter deck. In less than a quarter of an hour, the cable was cut and the vessel was forging ahead toward the northern headland of the harbor. The people on shore evidently expected an attack upon the town, as we inferred by the beating of drums, the sound of which we could hear distinctly. The flashing upon the battery led us momentarily to expect a salute from that quarter. We were almost out of range when the expected messenger came rushing over the water, soaring high above our mastheads, and dashing into the waves half a mile ahead of us. Ere another could be brought to bear, we were under cover of the merchant vessels anchored in the roadstead.

The satisfaction we felt now vented itself in three hearty cheers, and we began to think of looking after our prisoners. Three poor fellows were killed and we consigned them to the waves. Five others, badly wounded, we relieved for the present to the best of our ability. Of the remaining nine, two belonged to the Mexican custom-house, and others had formerly composed the crew of the vessel. The schooner proved to be the *Pelicano de Campeche*, and had been seized by customs authorities on the day of her arrival for having contraband goods on board. The process of seizure had been effected but about one hour before we appeared off the harbor. From the two custom-house officers, we learned that the character of our vessel was well known upon our first appearance, and thus everything had been done that was possible in so short a time to foil any attempt on our part to capture the vessel. The schooner had been partially stripped of her sails, and thirty-six soldiers, under the command of a lieutenant, had been put on board to defend her in case of attack. These, for the most part, had jumped overboard and swam to shore. They had tossed in the sea the corpses of three of their killed, to get them out of the way, while in the harbor, making the number of the enemy killed six and wounded five, as near as we could ascertain. On our side, with the exception of a few slight bruises and scratches, no one was hurt.

We continued making short tacks across the harbor until about 3 p.m. when the wind shifted off the land and we ran out, and at daylight came-to under the lee of the *Liberty*. In the hurry and confusion of getting the prize underway, one of our own boats had broken adrift and we had none left but the one belonging to the *Sabine*; and as it was impossible for us to do without one, we gave up all idea of restoring it. A spare suit of sails was instantly sent on board of the prize and bent, and the able-bodied prisoners were transferred to the *Liberty*. After the further transfer of sundry boatloads of flour, apples, butter, etc., to that vessel, not forgetting to mention some two or three thousand dollars in gold and silver coin, nautical and musical instruments, etc., we filled away, with instructions to make our way to the port of Matagorda without delay.

Under the American Flag at Matagorda

Among a crew composed of such materials as ours, there were, as might be expected, some strange characters—one, especially, in the person of a young man belonging to Liverpool, England. This queer specimen of humanity was an incorrigible ghost-seer, a fact of which we had all become well aware. Unfortunately for him, a few days before he joined the *Liberty*, an unhappy drayman who boarded at the same house with him in New Orleans, took the responsibility of shaking off this mortal coil by choking himself with a couple of yards of hemp. Although far removed from the scene of the transaction, our shipmate, who was usually called John the Drayman, was continually receiving visits from the spirit of his departed friend. Often in broad daylight, his eyes fixed upon vacancy, his whole frame shivering with mortal fear, his features of a deadly hue, and so completely unnerved as to be deprived of the least power to help himself.

At first, when these fits came upon him, the stoutest-hearted among the ship's company quailed beneath his unearthly appearance. But after his own explanation of the cause, attributing the effect to the appearance of the spirit, he became in equal degree an object of ridicule to the Hoosiers among us, and of anger to the old salts who, being strongly tinctured with superstition themselves, were but little pleased at the idea of a ghost's being upon terms of intimacy with anyone on board of the vessel. Therefore, upon the first symptoms of the approach of the unwelcome visitant to our mortal shipmate, a few buckets of water from the former and a series of kicks vigorously bestowed by the latter, had a strong tendency in most cases to dispel the illusion. In one instance this system of discipline came near having a fatal termination, as the miserable creature suddenly sprang overboard from his persecutors, and was with great difficulty rescued from a watery grave.

After having got our prize underway, as I have before related, Mr. Mayo directed this individual, who had formed one of the party, (for what purpose I know not, unless to get rid of him,) to light a fire and prepare some fresh coffee, of which there was plenty on board, for the refreshment of the men. In obedience to this order, the fire was soon made and ere long, the kettle of water sent forth its clouds of steam, to the satisfaction of all hands, and particularly of ghostly John who prided himself upon his culinary skill. Already the fragrant aroma of the crushed berry was regaling the noses of some half dozen shivering bipeds congregated at the galley door, impatient for the treat when, quick as a flash, the figure of our newly installed *cocinero* bolted from the galley, his hair standing on end and, with a loud shriek, measured his length upon the deck.

"What's the matter?" burst from the lips of the startled group, completely aghast at this sudden antic of their shipmate.

"Blast him!" said Mayo. "He has seen the ghost. Jump in there, one of you, and look after the coffee."

A lank Hoosier named Letcher, who also rejoiced in the soubriquet of "Beans" from his partiality to that vegetable, now entered the galley, but it was only to make a rapid transit through one door and out of the other, in a scarcely less frightened state than the first occupant. The men were now really alarmed. The daredevils who had but an hour and a half before rushed at the points of the bayonets leveled at them by the enemy now fell back from the galley with as much precipitation as if it had contained a mine of gunpowder about to be fired. "Well," cried Mayo, with an oath, "you are a bold set of fellows. What the hell is the matter with you? Stand back here and let's see what's the row." Pushing the men aside right and left, he made his way to the galley and, putting his head inside the door, instantly jerked it out again. At the same moment a most unearthly groan saluted our astonished ears.

We were completely nonplussed and stared at each other in a manner that would have amused an uninterested beholder, while John the Drayman whimpered out, "Why don't you go in *now*?" and our friend Beans gave it as his opinion that the devil was in the galley, swearing that he saw his eyes. We now drew towards the galley in a body, and when within a few paces, a succession of low moans fell upon our ears, which caused a few of the most superstitious to crowd back. Meanwhile, Mayo and the rest of us, convinced that the sounds were of an earthly nature, commenced a thorough search of the place, which resulted in the discovery of a tall cadaverous specimen of the genus homo, whose body and limbs were compressed into a space of scarcely sufficient capacity to have squeezed a small child.

Firmly fixed between the back of the stove and the caboose, he was lying on his side, his head and feet hidden from sight by a pile of wood upon each side of the range. His presence had not been discovered by any of the occupants of the place until his groans, extorted by the roasting process he was undergoing from the hot stove, betrayed him. Firmly believing that making himself known would have been but the first step to his being dispatched, he had remained in this uncomfortable berth until he was fairly roasted out. The poor fellow was almost dead, between fear and burning, and so much exhausted as to be utterly incapable of extricating himself, and he could not be released without getting burnt still more until the stove was moved. We were finally obliged to unlash the galley and raise it, when he rolled out upon the deck, as pitiable an object as I ever saw. He was the cook of the prize, and at the commencement of the attack had stowed himself away in this place for safety, from which he had put the whole of our victorious band to flight. Being an Italian, we received him into the service, and although he was no fighting man, in his line as a *cocinero* he afterwards rendered the state some service.

We parted with the *Liberty* on the night of the 5th, and saw no more of her until we joined her off Pass Cavallo at the entrance of Matagorda Bay. The vessel was a frail Mexican can-built craft of 200 tons burthen, and evidently intended for the contraband trade for there was hardly a spot about her which did not, upon strict search, disclose some secret receptacle for the concealment of smaller kinds of merchandise. By the manifest, a copy of which we found on board, her cargo purported to be five hundred barrels of flour, potatoes, apples and onions, together with some fifty half-barrels of butter, besides an almost endless variety of other articles of provision and merchandise, all of which made the cost value of the cargo somewhere about sixty thousand dollars. The run of the vessel and the forepeak beneath the forecastle floor was filled with the finest quality of chinaware and the private lockers were stowed with jewelry, musical instruments, rich dry goods, etc., most intended to be smuggled ashore. Seldom was so rich a cargo shipped on board a vessel of her class, and as these various articles were revealed one after another, our visions of prize money gradually expanded until the fortune of each had assumed a magnitude very gratifying to contemplate.

On the morning of the 9th of February we made the coast at the entrance of the Bay of Matagorda. Here we found the *Liberty*, in company with the brig *Matawamkeag*, having on board a company of volunteers under the command of Captain Morehead. These volunteers were from New York, and

I immediately recollected having spoken the same vessel with the same party on board, in the month of December previous, while making a short trip to supply the lighthouse at Key West with oil. They had been captured and detained at New Providence by the British authorities for committing some depredations in the way of supplying themselves with livestock, without troubling themselves about the payment for the same, and were detained about six weeks.

On coming in sight of our consort, we answered the private signal made by the *Liberty*, and in the course of an hour were at anchor alongside of that vessel. The pilot was already on board of the brig, awaiting the subsiding of the heavy swell which set in from the Gulf, rendering the passage of the bar dangerous. We came to anchor about 9 a.m. and were anxiously awaiting the moment when we should receive the signal to weigh, when two more sails were reported in the offing, standing in. These newcomers were soon made out to be two large vessels—the sloops-of-war *Boston* and *Natchez*.

As soon as their character was ascertained, their presence occasioned considerable uneasiness on the quarter deck, particularly when it was discovered that they still continued to stand in towards us. By the help of our glasses, we could discover that they were feeling their way with the lead with great caution. This boded no good to us, and while Walker was easing his mind with a volley of imprecations at our commander's stupidity, the *Liberty's* boat made its appearance. A few orders were instantly given and the boat returned. The next moment, all three vessels were simultaneously heaving short, and ere our sails were loosened, the spars of Uncle Sam's cruisers were stripped of everything but their three topsails, and they were clewed up, and came down by the run upon the caps as the ponderous anchors splashed from their bows, and they came-to, head to the wind.

We were underway by the time they had launched their boats, and the brig, with the pilot on board, took the lead, the *Liberty* following and the *Pelican* bringing up the rear. The two former passed the bar without difficulty. We alone, in the prize, were destined to be unfortunate. When about the top of the bar, we struck heavily, completely stopping our headway, at the same time pressing us to leeward. Our situation was now desperate, as no assistance could reach us in time to be of service. The vessel was every moment lifted upon huge rollers, which on passing would let her drop down upon the hard sand with fearful force. We now broached to with our broadside to the sea, and soon struck with such force that the masts went by the board. As Walker shouted a precautionary "Look out for yourselves!" to the men, the vessel rolled partially over, striking upon her beam ends, when the whole fabric broke up in an instant, and I found myself struggling in the waves.

My brain was in such a whirl of excitement from the events which had passed so rapidly (for five minutes had hardly elapsed from the moment we struck until the destruction of the vessel,) that I had no opportunity to think about what course to pursue. It seemed to me as if it was in a dream. Once, and once only, I grasped at something, and got hold of and touched it; but it eluded my grasp, and after that I recollect nothing, until I struck the sand with such force as to awaken me to a sense of my situation. I opened my eyes just in time to find that the undertow was again taking me out to sea and, although I grasped at the sand and made giant efforts to gain a foothold, I felt it was all in vain. I had almost resigned myself to my fate when my course was arrested by the grasp of a friendly hand on my arm, and the voice of Jim O'Connor fell upon my ear with, "Ye are not going to lave us that way, are yez?"

What happened during the next hour I know not. When I recovered my consciousness I was lying upon a sandhill in the warm sun. Three or four others of the crew were lying beside me and the rest, assisted by a strong party from the *Liberty*, were busy securing whatever came ashore from the wreck. In a short time I felt I had sufficiently recovered to lend a hand, and by nightfall we had succeeded in saving upwards of two hundred barrels of flour in a damaged state. Many of these were stove, and some which were knocked to pieces were found to contain in the center a 25-pound keg of powder, which we afterwards found was the case with all of them. We likewise saved about two thousand dollars worth of jewelry, concealed in false bottoms of thin trunks made of mahogany veneers, which were otherwise filled with dry goods, most of which had burst out.

Thus in a moment, as it were, vanished our daydreams of wealth, and it was natural we should curse the unwelcome strangers who had so signally destroyed the success of our cruise at a moment when it appeared to be complete. The various articles saved were piled up out of reach of the surf, and at night the majority of our party rejoined the *Liberty*, which lay near the pilot-house. The *Boston* and *Natchez*, after witnessing the mischief they had done, hove up their anchors and took their departure, as if perfectly satisfied with the result of their operations, and we saw them no more.

After a short sojourn at this place, everything being secured from the wreck, we weighed our anchor and proceeded to Matagorda, which lies some forty miles north of Pass Cavallo, at the head of the bay. It was a bright morning, the breeze blowing fresh and well aft. The mainboom was swung well forward by means of the boom-tackle and pennant. The peak down-haul having got loose, it swung out to the end of the spar, and Mr. Mayo ordered our friend, John the Drayman, who had escaped with the rest of the wreck, to lay out and bring it in. In obedience to the order, the unfortunate man proceeded to the end of the boom, laid hold of the rope and had returned about halfway when he was seen to spring to his feet upon the narrow spar and, throwing up his arms, his eyes bulging from his head, he shouted, "He is coming!" At that instant, he fell from the boom, the water receiving and closing over him forever. The vessel was brought to the wind and the boat lowered, but nothing more was seen of John the Drayman.

At half-past one, we came in sight of the anchorage, nine miles below the town of Matagorda. Here we discovered three other vessels at anchor, which proved to be the Texan schooners-of-war *Independence,* of seven guns, bearing the broad pennant of Commodore Charles E. Hawkins; the *Brutus,* of eight guns, Captain Hurd; and the brig *Durango,* merchantman. These vessels of war, although belonging to the Texan service, carried the flag of the United States. On coming in sight of them, our vessel immediately showed her number and made the private signal prescribed by the rules of the service, but these gallant seamen appeared to take no notice. Worse than this, our glasses disclosed to us that they were clearing their decks for action. This was no sooner discovered than our drum beat to quarters, and in a few moments our decks were cleared, the guns cast loose, matches lit and every preparation made to meet the strangers either as friends or foes.

We were now fast approaching the anchorage. The men, ranged around the guns, armed to the teeth, apparently looked upon the prospect of a brush with our friends with as much pleasure as they would with the Mexicans. When within hailing distance, an officer from the deck of the Commodore demanded the name and character of our vessel. To these questions, Captain Brown replied interrogatively, "Don't you see the number at my masthead?" This answer did not suit his excellency, the Commodore, and the questions were again repeated, accompanied by the threat of firing into us in case of not receiving an explicit answer. The threat failed at producing the desired effect, as the matches were instantly snatched from the tubs where they were burning, while our commander roared in a loud tone of defiance, "Fire, if you dare!" This answer, and the display of our warlike attitude, in connection with the ugly-looking set of customers upon our deck, cooled the courage of his excellency amazingly. The drums of both vessels beat the retreat from quarters, and passing between them unmolested, we came to anchor ahead and to windward of both of them.

In the course of a couple of hours, we received a visit from the Commodore, a medium-sized gentleman with a very severe expression of countenance, dressed in the full uniform of a post-captain of the United States service. He was accompanied by his flag lieutenant, likewise in full uniform. Their appearance, taken in connection with the fact that the flag of the United States was unlawfully waving over the vessels they commanded, which vessels belonged to no recognized power, presented all the points of a piratical masquerade, which our own appearance helped to make up. The Commodore, well skilled in naval etiquette, bowed gracefully as he reached the gangway, and proceeding to the quarter deck, immediately gave directions—or, I should rather say, orders—to haul down the Mexican flag, which was flying at our peak, and to substitute the American flag in its place, alleging that the Mexican flag was not being recognized on account of the figures before mentioned, and we would be subject to capture. He gave our commander to understand that no umbrage would be taken by the American government at our using its flag until the independence of the country was declared, which measure was already in progress, and when accomplished, a flag would be at once adopted and presented to the world for recognition. To all this our commander replied that it would do very well for those who designed to lie at anchor in the harbors of the country and eat the government provisions, without rendering any service in return. This answer seemed to exasperate his excellency somewhat, as a very exciting discussion immediately commenced, which was carried on for some time with considerable warmth. The controversy appeared to have arisen from the refusal of our commander to recognize the authority of the Commodore over the *Liberty*. At length, that officer took his leave, turning and delivering himself with "The Lord be with you!" as he mounted the gangway. This was the only time he honored the vessel with his presence during his stay in the port.

The next day, the boat of the *Independence* came alongside with an armed crew, under the command of the flag-lieutenant, bringing with him a new commander for our vessel in the person of Captain Wheelwright. Measures had been taken by Captain Brown for his reception, and he was met at the gangway and escorted to the cabin, where the dangerous documents in his possession were taken from him before being opened. [NOTE: The newly formed Texian government officially commissioned George Wheelwright to command the *Liberty* on March 12, 1836. Cushing's dates of this Matagorda rendezvous, as such, may be incorrect, since he has it taking place in February. If command did pass to Wheelwright from William Brown, it was only temporary, as Cushing has Brown back in command again in just a few pages. It is known that charges were brought against Brown and Wheelwright *did* assume command, but this did not occur until the summer of 1836.] While this business was in progress aft, the men were piped to grog and the boat's crew of the *Independence* were invited on board to partake with them. The unsuspecting midshipman in charge gave the necessary permission, and as they came on board, their places were instantly filled, to the astonishment of the middy, by a number of our marines who, without even saying "by your leave," immediately passed the arms of the boat's crew on board, where they were safely transferred to our arm-chests. The lieutenant and his party were then politely dismissed with the caution to make no such attempts in the future.

Captain Wheelwright remained with us, a prisoner on board of the vessel which he had received orders, from the only legal authorities in the country, to assume command of. No further communication was held with us by the Commodore and his coadjutors, and we busied ourselves about our own affairs, taking no more notice of them. It could have been nothing but arrant cowardice that prevented them from attacking us, as either of their vessels was heavy enough to have blown us out of the water. And although they both brought forward the plea that they were short-handed, the two vessels could have mustered some eighty men, which doubled the number on board our vessel. It was a significant fact that neither of these officers, Hawkins and Hurd, ever made any demonstration against the enemy during the war. Their vessels were ever found safely at anchor in some of the different harbors of Texas, while the others, under captains Jerry Brown, Hoyt, and Thompson, of Mexican celebrity, were always found harassing the coast of Mexico and paying them in their own coin for the devastations they had committed in our country.

For several days after these events transpired, nothing of note occurred. The men lounged about in listless idleness, and the time began to wear heavily upon us, when we received the startling news that our army had fallen back from the Colorado River and that the Mexicans were advancing upon Matagorda. Upon the receipt of this information, the *Independence* and *Brutus* got underway and sailed for Galveston, leaving us and the *Durango* the sole occupants of the bay. Nothing could have pleased us more. From the commander down, all expressed their gratification at this unlooked-for movement and it was at once resolved to visit Matagorda and render the flying inhabitants all the assistance that lay in our power.

To ascertain the state of affairs in the town, which we had not as yet visited, the captain's boat was at once manned and proceeded to the place. My duty as midshipman gave me the opportunity to participate in every boat expedition, and I accompanied the captain in all cases while on board of our vessel. We were likewise accompanied by Colonel John A. Wharton, an officer in the army, invalided in consequence of a wound received at the siege of the Alamo. [NOTE: Wharton actually received the injury that lamed his right hand during a duel with William T. Austin.] After a long pull against the strong current setting out from the Colorado, we reached the landing of the town, which is situated upon the right bank of the Colorado where it empties into the bay. The landing, a large space cleared of the heavy canebrake which everywhere else lined the bank and extended into the river some forty feet, was literally covered with piles of merchandise of every description. It had been removed from the town in readiness for shipment on board of whatever vessels might chance to enter the port. But the premature intelligence that the enemy was near the place had put the people to flight and they'd left their property to take care of itself.

Leaving the boat in charge of the men, the captain and Colonel Wharton proceeded to the settlement, some three hundred yards from the top of the bank, and I was ordered to accompany them. A small white flag was flying from the gable of a large wooden storehouse which stood nearest to the river. This I was commanded to take down and once that was accomplished, we continued on. The town, consisting of some twenty or thirty houses and stores, was completely deserted. Not a soul was to be seen in any direction. We entered a store and found the counters piled with dry goods, clothing, etc. A characteristic feature of the habits of the country struck me forcibly, as my eye fell upon the immense quantities of playing cards which lay in every direction.

From this store, we proceeded to the place which had formerly been the hotel of the town. Here we found a superannuated negress who gave the massas to understand that she had been left in charge of the hostel while massa and missus had sought safety beyond the Brazos. We immediately installed ourselves in this house and, after ordering dinner, the materials of which the hostess promised to provide, the subject of a plan of operations was brought forward. It was outrageous to let so much property as was in this place fall into the hands of the Mexicans, while there was even a chance of saving it. But the uncertainty of the whereabouts of these gentry puzzled us in the extreme. At length the wise course was adopted, of acting for the present as the absolute necessities of the moment required. By the time this conclusion was arrived at, our hostess gave notice that dinner was ready. This disposed of, I was ordered to return to the *Liberty*, with orders for Mr. Walker to send immediately on shore as many men as could be spared from the vessel, together with all the boats that could be procured. Orders were also given to have the brig *Durango*, which was of light draught, sent as near the place as possible.

With the injunction to use all possible dispatch, I started upon my mission and in a few moments reached the boat. Here, a scene disclosed itself which rendered abortive every effort on my part to execute the orders of my superior. The men were stretched upon boxes and bales, and otherwise variously bestowed—some dozing, others wrangling and fighting, and all of them so drunk that any attempt to proceed for the present would have been worse than useless. I now for the first time noticed some twenty or thirty barrels and quarter-pipes of liquors, which were intermingled among the rest of the merchandise. A number of large tin pans were scattered around, partly filled with villainous brandy, which I immediately poured upon the ground. Proceeding to the house, I reported the state of things at the landing to my commander, who was by this time beginning to feel the effects of the copious libations he had indulged in, probably to promote digestion. After receiving the communication with drunken gravity, he ordered me to bring them up to the house and keep them by me until they became sober. Returning to the landing, I succeeded in coaxing a couple of the rascals to the house where, as the only means of securing them, I gave them as much liquor as they could drink.

The day, the early part of which had been fine and warm, was now just the reverse. The rain fell in frequent and heavy showers, and by 4 p.m. it rained in torrents without intermission. In a short time, the remainder of the boat's crew, partially sobered by the cold bath which aroused them from their drunken slumbers, came staggering along one by one, and by the time night closed in, we were all snugly housed.

An immense fire in an out-house used as a kitchen served to dry the clothing of the men who, after they had partaken of a warm supper, began to hanker after the siren of the still, and ere I was aware of the maneuver, one of their number was dispatched to the bank of the river, armed with an old tea kettle, who presently returned fully charged with the extract of rye. I could have taken possession of the liquor, and I had already received orders to prevent them from procuring any more, but this order was more easily given than executed and, knowing that my commander's situation enabled me to do so with impunity, I chose to disobey it. By ten o'clock, the men were all drunk and snoring.

After destroying the liquor, except just enough to give each of them a glass in the morning, I fastened the doors (windows there were none) and took possession of a fine featherbed, which the old negress, into whose good graces I had already got, had assigned me. After listening a short time to the loud brawling of my respected chief and his companion, who were making a night of it in an adjoining room, I fell asleep—the only sober person in the house, with the exception of our colored landlady.

During the night I was aroused from my slumbers by something which felt like a coarse rasp grating down the entire length of one of my legs, almost stripping off the skin. Quickly raising and turning myself, I discovered that some person, with his clothes completely saturated with water, had introduced himself between the sheets and was already sound asleep. In endeavoring to make out who it was, my fingers encountered the gold lace shoulder straps of the coat of Colonel Walker. A little further examination revealed the fact that he had come to bed like a trooper's horse, all standing, without omitting his boots. The colonel evidently had an eye to a hasty retreat, in case of a surprise by the enemy. My compatriot was an uneasy bedfellow, and it was with a great deal of satisfaction that I beheld the advent of another day.

As soon as it was light enough, I parted company with my illustrious bedfellow and, dressing myself, proceeded to reconnoitre. My commander was snoring in the ashes of a large fireplace in another of the out-houses, his clothes drenched with rain, a large puddle of which had drained from his clothes and lay beside him. The old negress, stooping above him with a lighted match, was trying either to set his nose or the pile of wet wood in a blaze; and from appearances, the captain's nose would have burnt much more freely than the material of which she was trying to make a fire. The two officers, it appeared, had been upon some nocturnal excursion, for what purpose I could not ascertain. The out-house where the boat's crew were held in durance gave forth no sounds except the loud snores of the men.

Everything being safe in that direction, I proceeded to the landing and found the boat half full of water. I baled it out and, having got everything in readiness for a start, returned to the house. Here I encountered a tall sun-burnt man, dressed in a buckskin rifle-shirt and breeches, fringed at the seams and edges. A long rifle was slung under his arm, with powderhorn and bullet-pouch to match. A coonskin cap, Indian moccasins, a heavy bowie knife and pistols completed his equipments and indicated the character of his business in language plainer than words. He was mounted upon one of the sturdy native horses, much above the usual size of the Mustang breed, which had the appearance of having been hardly ridden; but the vicious nature of the race still gleamed from the wild eye of the animal, whose mouth unwillingly held one of the heavy Spanish bits, furnished with rollers to prevent his taking the reins of government, as well as of the bridle, into his own control.

Half a glance told me that this person was one of the scouts belonging to the army. Disengaging his feet from the stirrups and springing to the ground, he extended his broad palm and, giving my hand a hearty shake, saluted me with, "Well, stranger, how fare ye in old Mata? And where's the garrison?" After assuring him that the garrison and all in the house were well, I inquired if there were any of the enemy near the place. "Well yes, stranger, there's a right smart chance of serpents not far off, but I guess they'll not be here right away. Old Filisola don't move very fast, though Santa's pushing on after old Sam like hell." All of this was perfectly intelligible and the appearance of

Captains Brown and Wharton put an end to any further questions on my part. After hobbling his horse, an operation performed by fastening the forelegs of the animal with a short trap to prevent his straying, the trio disappeared into the house, while I was directed to proceed to the *Liberty* with orders of the previous day.

Upon going to the out-house where the men were lodged, what was my dismay to find the whole front of it in ruins, and the birds flown. In an instant, I went to the landing where I found the crew already at work upon the casks of liquor. Almost at the risk of my life, I instantly upset the various vessels of liquor which had already been drawn and succeeded, after considerable expostulation, in getting the men into the boat and shoving off. Everything now seemed to promise well, and for the first half hour we made rapid progress toward the vessels. The men, upon leaving the river, had filled a couple of tin pans with water to drink when we should get into the bay, and now having reached a point where the water was rather brackish, the thirst became general. Every few minutes, one of them would cockbill his oar and take a drink out of one of the pans. Not suspecting any trick, I paid no attention to these movements until we had gone about two-thirds of the distance, when the strange actions of some of the men attracted my attention. In a few moments I was conscious of being sold—duped. Jim O'Connor was just in the act of taking a swig when I seized the vessel and my suspicions were confirmed. The water taken from the river before my eyes was a mere ruse. The men had been drinking whiskey the whole time! I now commenced a search in which I found the water untouched, while the men had succeeded in disposing of nearly a gallon of whiskey.

Anxious to get the boat as near the vessel as possible before the natural consequences of this proceeding should render me helpless, I urged the men to give way with a will, and inclined the course of the boat toward the north shore or head of the bay, knowing that if the crew gave out, the current, which set down toward the schooner, would carry me within hail. In a few minutes, the last act of the play was commenced by two of the crew striking their oars foul and pitching backward into the bottom of the boat, where they remained without even an effort to rise. With great difficulty I succeeded in saving the oars, and once more directed the boat toward the vessel. By the time I arrived within hail, there were but two men who could keep their seats, and they were nearly helpless. Walker observed the predicament I was in and sent a boat with sober men from the vessel to my relief. I at length reached my destination, where the officer in command, after hearing of the trick which had been played upon me, was not in very good humor and cursed me for a stupid greenhorn for allowing myself to be outwitted, at the same time treating the boat's crew to an unmerciful ropes-ending.

I was allowed to remain on board just long enough to get my breakfast, when I was ordered to return to the town with the captain's gig, while Mr. Mayo, with some twenty of the crew, followed with the heavier boats. Upon landing again, I found the party there augmented by the addition of a Mr. C., a gentleman who was the proprietor of a large plantation upon the Colorado. This person was accompanied by his daughter; and a beautiful girl was Miss Caroline, and mischievous withal, though not designedly so, as the hearts of the gallant doctor and purser of the *Liberty* will acknowledge. While she remained with us, she was the cynosure of all eyes—the toasted reigning belle of the day.

The men were instantly set to work at loading three large clinker-built lighters, which had been hidden somewhere in the canebrake, and I was introduced to the young lady as her escort, with orders to do her behests in all things lawful, by our commander, who I had hardly supposed possessed so much gallantry as he exhibited on this occasion.

The place now presented quite a contrast to its appearance on the morning of our first landing; and as the fear of the immediate occupation of the town by the Mexicans wore off, we passed our time very pleasantly. The gentlemen busied themselves in facilitating the shipment of the merchandise, and the scout had again taken himself off to the prairies; while my charge and her humble servant, the only persons of leisure, went and came as we chose, seeking our own amusement. My fair friend, although somewhat older than myself (four years, I believe,) was an excellent companion. She had been reared in the lap of luxury, and had received an elegant education. She had accompanied her father through the United States in various speculations, and they purchased a large tract of land upon the banks of the Colorado. They had settled themselves down for life, they thought, but scarcely four years had elapsed when the approach of the Mexicans drove them from their happy home, which I am sorry to say they found utterly ruined on their return.

Day after day passed, and the most valuable portion of the goods in the place were safely transferred to the vessels, and I became conscious that my very agreeable duty was fast drawing to a close. But there was no help for it, and I endeavored to put a good face upon the matter. It was just at sundown, and the heavily laden lighters had but a few moments before left the landing when our friend, the scout, galloped up to the house. Calling out Captain Brown and the Colonel, he spoke a few words, the purport of which the distance prevented any understanding, and then hurried off. This, from the agitated manner of the messenger, I conjectured to be a warning that the Mexicans were near at hand. The surmise was correct. The captain almost immediately entered the room, and directed me to have the gig manned at once. As I left the house upon this duty, I observed Colonel Wharton mount Mr. C's horse, and taking the same course as the scout had previously done, was soon lost to view. In five minutes I reported the boat ready. "That will do. Away with you, (addressing the young lady at the same time,) away with you to the *Liberty*."

Hurrying my charge along, I soon seated her in the boat, and in a few minutes we had cleared the river and entered upon the broad waters of the bay. Many were the questions put to me by Miss Caroline, ere we reached the vessel, in regard to the place that was assigned her on board—a matter in which I was as much mystified as she was. However, I managed to satisfy her as well as I could, though I feared that her residence on board would not be as pleasant as it had been at Matagorda, poor as it was there. In this I was mistaken, though, for in three days she had accommodated herself so well to the situation in which she was placed as to be perfectly contented. She had never been upon the water before in a small boat, and the motion of the gig alarmed her. We got safe on board, however, and my fair friend was safely installed in the cabin. I found the *Liberty*, as well as the brig, had been loaded as heavily as possible. In fact, had we been attacked, our guns would have been useless until we had thrown overboard a large amount of property.

UNDER THE FIRST TEXIAN FLAG

The captain came on board an hour after I had reached the vessel. My duties being over for the day, and feeling somewhat weary, I turned in, and had already in my dreams got back to the town with my late companion when I was roughly shaken by the shoulders and I awoke. It was Walker.

"What is it?" I inquired, half asleep.

"Boat service—want to go?" I answered by springing out of the berth. "Turn in again, boy," said Walker, laughing. "Not yet. I'll call you when we are ready."

Before I turned in again I went upon deck, and the first thing I saw by the light of the moon was some half dozen of the watch passing a quantity of small arms into the heaviest of the lighters. The boat pulled sixteen oars. I went below again, and passing to my resting place, a hand from behind the curtain of a berth beckoned to me. I stepped up to the place and a voice whispered, barely audible, "Don't you accompany Mr. Walker. They are going to fight the Mexicans." I made no promise, as I could not think of Walker going anywhere without my being with him. True to his word, Walker roused me at about two in the morning, and bidding me to hurry up, went upon deck where I joined him in a moment afterwards.

Everything was ready, and after receiving orders to fire the place and, if possible, bring away a couple of ship's guns (six-pounders, which stood at the side of the road leading from the landing to the town) we put off on our enterprise. We had the advantage of a fair tide, which, the boat being large and heavy, enabled us to get along very expeditiously. Very soon after starting, Walker cautioned me to mind the bearings in case of fog, and then rolled himself up in his watch-cloak and went to sleep.

When within a mile of the mouth of the bayou, the anticipated fog made its appearance, and in a few moments became so dense as to prevent me from seeing the bow of the lighter. I aroused my superior and we kept on an hour longer. We'd begun to fear that we had miscalculated our distance when the boat forced her way through the crushing cane into a large brake. Not knowing the precise locality, the oars were trailed alongside, and making fast to the cane, we awaited the approach of dawn. It was not long before the increasing light put all hands on the alert to discover our whereabouts, but the dense fog prevented satisfactory result. Suddenly, the rattling of some dozen or more drums, far below where we were, fell with startling distinctness upon our ears.

"What the fury is this?" muttered Walker. "Here's a pretty kettle of fish! We've got ourselves into a hornet's nest! Boys, have your arms ready. We shall be lucky if we get out of this scrape without hard knocks. One thing, at all events, I want you to understand: there is no mercy to be expected by anyone who falls into their hands and I, for one, shall never be guilty of the folly of trusting them." In these views we all heartily concurred, and extricated the boat from the brake with all possible dispatch.

The fog now became gilded by the rays of the rising sun, and presently rolled itself up like some vast curtain, revealing the unwelcome assurance that if we escaped our present peril, it was only to be done by running the gauntlet within pistol-shot of some six or seven hundred Mexicans posted on the top of the bank at the landing. This was in itself enough. And fearing their numbers might be increased, as the sound of drums and trumpets came from every direction, the men gave way at the oars and the boat dashed down toward the landing.

As a ruse and at Walker's direction, I had affixed a small white flag in the bow of the boat, and stood with boat-hook in hand as if ready to prevent the too forcible contact of the boat with the shore. We now approached the landing, which was occupied by squadrons of officers, whom we had not discovered before. "Now, my lads, mind what I say and when I give the word, pull for your lives," said Walker. The shore was lined with Mexican officers, a number of whom had affixed their white handkerchiefs to their swords, which they waved in answer to our flag of truce, with the invitation, "*Salta para tierra, muchachos nosotros somos amigos.*"

"No doubt, you bloody cut-throats!" muttered Walker, rising and returning their compliments. The boat's bow was now turned obliquely toward the shore, and while we were seemingly striving to land her, the strong current of the river, acting upon her broadside, swept her rapidly down toward the brake at the lower point of the cleared space. The men, although appearing to work hard, were only playing 'possum, in accordance with Walker's whispered directions, to which the men on shore began to open their eyes. They at once assumed a threatening attitude, but it was too late, for we were almost under cover of the brake. As we were swept past, the bow was kept pointed for the landing and was now fairly upstream, her headway being completely stopped. "Hurrah, my lads! Stern all!" shouted Walker, instantly jerking the rudder from the gudgeons.

With a couple of backstrokes vigorously given, our boat glided under cover of the friendly canebrake and we were safe. A shower of balls, altogether too high for their intended object, whistled over our heads and found a resting place in the waves fifty feet beyond us. In derision, we pulled out where we could command a view of the long line of troops on the bank and, seizing our arms, we returned the enemy's salute. With three hearty cheers for our success, we once more turned our boat's head for the vessel, on board of which we arrived without any further adventure.

As it was no more than reasonable to suppose that the enemy would have some naval force to cooperate with them, our commander concluded that the sooner we were out of the bay, the better. Accordingly, the dawn of the next day found us underway, in company with the brig, for the Pass Cavallo, where we came to anchor towards night. From here, while preparations were being made for sea, I accompanied an expedition to burn some government storehouses at Cox's Point.

When I returned on board, I was proffered and accepted the command of one of the lighters which, loaded with flour, was to be taken round to Galveston. Having received permission to select my own crew, I chose my friendly Irishman, Jim O'Connor, and three others, and at once took command of my charge; not, however, without taking leave of my fair friend, Miss Caroline, who expressed her regret at our being obliged to separate, and gave me to understand that she felt much annoyed by the fulsome flatteries of certain of the quarter deck gentry on board, to whom I was aware I was indebted for my present command.

Everything being in readiness for sailing, the news was revealed that the independence of the country had been declared. A flag was placed in my hands, with directions to fasten it to the halliards and hoist it. This was done, a salute of three times three from all hands greeting the first unfolding of the flag to the breeze. This was the first Texan flag that had been hoisted on board any of our national vessels, and differed somewhat from the one afterwards adopted. It was said to have been the one used at the siege of San Antonio de Bexar the December previous—a blue field, and in the upper corner a single star of five points, over which was stretched a blood-red arm and hand, holding a naked sword; the rest, red and white stripes alternately. The center white stripe contained the word "Independence" in large letters reaching its whole length. At the close of this ceremony, another, in which I was more intimately concerned, occurred. Through the hands of my sweet friend, I received from Captain Brown the present of a beautiful dirk, with belt, etc., accompanied with many gracious words from the young lady, by permission of the discretionary power delegated by the commander on the occasion.

I was now ordered to proceed on my voyage, and in a short time, in company with the *Liberty* and *Durango*, we had passed the bar, and gaining a good offing, shaped our course for Galveston. My little bark made much slower progress than the larger vessels, and by sunset they were nearly hull-down ahead. But I had been given to understand that I would have to depend on my own wits, and therefore had made special preparations for every emergency. The nights, especially in the winter and spring, were very chilly. I had prepared everything for a liberal supply of hot coffee and had not forgotten to provide a large jug of whiskey, which latter article, however, I kept a good lookout for, serving it out to my company according to the rules of the service—a measure of precaution which did not fail to produce a little grumbling. Everything went on smoothly for the time being. The men munched their salt junk and biscuits, washing it down with copious libations of coffee, and amused themselves with singing songs and spinning yarns until the lights of the vessels, which we had occasionally seen, disappeared altogether. The wind blew rather too fresh for our comfort, and our little craft occasionally shipped more water than suited us. But we managed to keep her free by means of a couple of buckets, which were at length kept constantly going.

But the worst luck was yet to come. A little after midnight, the wind shifted to the north and obliged us to stand off shore. This was unpleasant; however, we made ourselves as comfortable as circumstances would allow. The wind soon after moderated and we were relieved from any further necessity of baling. Toward morning, the wind again breezed up, and we made rapid progress during the whole day, nothing being in sight but the low range of sandhills which fringed the coast. Long ere sunset, the boat's crew had sang their songs and spun their yarns for the twentieth time, and at length had relapsed into a drowsy silence, unbroken save by some exclamation of impatience at the decline of the wind, which now threatened to leave us.

After a calm comes a storm, and the appearance of the weather foretold there was to be no exception to the rule in our favor. Heavy banks of clouds began to gather in the north and east. But, as if to reassure us, the breeze again set in from the southeast, and we bowled along merrily through the smooth water. The night was fine and the azure vault above us was studded with myriads of stars, but the clouds before mentioned still held their place, as if awaiting their turn at the bellows. About ten o'clock I was obliged, by the great increase of wind, to reef the mainsail; and taking in the foresail, I kept on my course for an hour longer. By this time, the sea had become quite heavy and, the wind blowing still harder, we several times narrowly escaped capsizing.

We were thus so much occupied with our own affairs that we had not noticed the approach of a large schooner, now within hail, and on board of which we were already discovered. I recognized the vessel at once as the *Invincible*, under the command of Captain Jerry Brown, a brother of my own commander, who immediately hailed us to know how we got along. He advised me to beach the boat as soon as possible, as there was every appearance of a norther. By the time he had imparted this information, his vessel had forged out ahead of further hearing and presently filled away, leaving us to chew the cud of bitter fancies which he had prescribed.

To make any further attempt to keep on my course I knew would be next to madness, but the dangerous alternative was equally as bad. As the latter offered the only chance of escape, I resolved to make the attempt. In this maneuver, the boat education I had received at Cape Cod was of great service to me. While living at Wellfleet, I had frequently accompanied the hardy fishermen of the place to the outside coast of the cape, in their quest of sea bass, large quantities of which were formerly caught there. The surf breaks heavily upon the coast and the people are very expert in the operation of beaching their boats. I would much rather have trusted to a good set of oars, but I had not the choice, so I commenced operations by throwing overboard the top tier of barrels of flour. This measure relieved the boat so much that she ceased to ship water to any great extent and road the huge swells handsomely.

We were now rapidly approaching the coast, but the immense fields of white caps that intervened, and which we could perceive as we arose upon the crest of the waves, combing, bursting and towering high in the air, almost appalled me. But I kept my own counsel, although I felt my heart already many degrees below zero. Determined to give the boat all the assistance in my power, I shook out the reef in the mainsail and hoisted the foresail. As she passed through the trough of the seas, where for a moment we were becalmed, I was fearful the returning waves would render the rudder useless; I had therefore substituted an oar in its place. The men pulled with a will, as each swell passed, keeping as much of the headway as possible, but the most fearful part was when the little craft, lifted astern by the seas, hurried along with such amazing velocity as to almost take our breath from us.

As we drew near the outer bar all was a wild chaos of bursting waters. How we got through, I know not; but we appeared to be passing through the air instead of over the surface of the water, and for an instant we found ourselves in much smoother water. Our sails now again became of service to us, and under their impulse we dashed on toward the coast, upon which the surf broke with a deafening roar. In compliance with the wishes of the men, who were wet and cold, the jug of whiskey was passed forward. After a hearty pull at the exciting beverage, they again stationed themselves at the oars for the final struggle, which was now fast approaching.

We were already within the influence of the outer breakers and were being at one moment hurried along at a tremendous rate, and then, as if by some counteracting power, as speedily checked in our mad career. Each succeeding wave increased the velocity with which we approached the shore, which soon brought our hazardous enterprise to a termination; for at length, rising upon the crest of an enormous roller, we were borne high upon the beach, where the boat was dashed upon the shore a complete wreck, pitching me and my companions headlong upon the sand. This happily terminated our voyage without any further injury to any of us than bestowing upon all a sound ducking, which, in view of all the circumstances, we were content to put up with. Upon examination, we found the barrels of flour stove so badly as to render their contents entirely useless.

Not knowing exactly where we were, I determined to keep the beach, as its smooth, hard, sandy road afforded greater facilities for traveling. The jug of whiskey was the only thing we recovered from the wreck uninjured, and Jim O'Connor took it under his patronage, carrying the article while it contained a drop, with as much care as a mother would her infant. The first thing to be done was to make our way to the sandhills and see what discoveries could be made inland. This meeting with the assent of all hands, we started, and in a few moments reached the top of one of the hills, which commanded a view of the prairie. From this spot, the sight of a long range of fires attracted our attention, to the great joy of my companions, who at once pronounced in favor of proceeding to them. Scarcely believing the enemy had made such progress, I was about to comply, when my ear detected a voice in the distance. I obtained silence for a second, and made out the Mexican *"Alerta esta,"* with a certainty which there was no gainsaying. This convinced us it would not do to proceed in that direction. Again returning to the hard sand, we used our utmost endeavors to put as much distance as possible between us and the dangerous locality before daybreak.

The morning light revealed no signs of either friends or foes, and we kept on our way. Though wet cold and hungry, and ignorant of distance or the direction in which most readily to seek relief, I concluded that the place where we had seen the campfires must be the mouth of the Brazos. In this surmise I was not mistaken, but the fires proved to have been on the opposite side of the river. About half-past two, we came to an inlet which separated us from the island. A broad sheet of water spreading out between it and the mainland convinced me that this was the west pass of Galveston Bay, which I had been told was for the most part fordable. Jim O'Connor volunteered to try the passage first, and for the greater security of the attempt, we constructed a small raft, fastening it together with our handkerchiefs, belts and the halyards of the boat's sails, which we had fortunately brought with us. We found no difficulty in procuring wood, the beach being heaped with immense quantities of every size, from the stick of firewood to the trunks of enormous trees.

Having completed our raft and launched it, we found that it would easily bear the whole of our party. So, having provided ourselves with good setting-poles and stripping off all the clothing we could dispense without freezing, we put off. The wind had already shifted to the northwest, but contrary to all our expectations, it hardly amounted to a stiff whole-sail breeze, yet was still enough to make the atmosphere very chilly. Our raft with its heavy load frequently grounding, those who could swim stripped altogether and waded, pushing the catamaran before them. We now got along finely and found ourselves fast approaching the channel, which we crossed without difficulty, and shortly after landed upon the island.

Resuming our clothes and pulling the raft to pieces, we proceeded to examine a hut which stood at the top of a beach. Everything in and around the place showed that it had been very recently inhabited, and that the owner had not as yet relinquished his proprietorship. A number of sides of bacon hung against the wall, a welcome sight to our famishing party. A few moments' further search disclosed a half-barrel of flour, some coffee, sugar, and what was more grateful to my companions, a small keg in which still remained a couple of gallons of whiskey. But all were too hungry, for the time being, to think of anything but the eatables, and in a short time a bountiful repast was spread upon the rough table of the shanty, to which we all did ample justice.

Taking Leave of the Liberty

The day was now spent and the men, having found an old slush lamp and a greasy pack of cards, took possession of the aforementioned whiskey and proposed to make a night of it. Looking upon the matters as a sort of thanksgiving frolic for our recent escape, I joined the party for the purpose of making away with the liquor as soon as possible. They had followed my counsels in everything since we had landed, but I was aware that the moment they became drunk, I should have an ugly set to deal with.

The greater part of the night was consumed in card playing, drinking, singing and uproar, but towards morning, the noisy tempest was stilled. I would have destroyed the element of mischief at once but to my surprise, on examination, I found the keg empty and all my endeavors to discover the whereabouts of the liquor were fruitless. Vexed at being thus baffled, I lay down upon the table and, following the example of the rest, was soon asleep.

I awoke shivering with cold, and after a run upon the beach to warm myself, I returned to the shanty and built a fire, for the purpose of making some coffee. The coffee pot or kettle had disappeared, and I at once knew what to search for to find the liquor, and examined the sand which formed the floor and scraped at it in every direction with my fingers, but in vain. I at length noticed that O'Connor, in his sleep, had rolled from the place where he had first laid. Here I sought and in a few seconds, the object of my anxiety was in my possession. Pouring three quarters of its contents upon the sand, I returned the vessel to its hiding place. Smoothing over the sand again, I left the hut and mounted one of the sandhills to see what discoveries I could make seaward. A couple of vessels far out upon the

Gulf was all that broke the monotony of the waves. Upon the prairie, at a distance of some six hundred yards, a flock of deer, which had been grazing and were disturbed by my appearance, tossed their graceful heads and, snuffing the air, bounded off with the fleetness of the wind, from what they no doubt considered a dangerous proximity.

In my ramble, I discovered a camp among the sandhills, where a low rack had been built on which was laid a large quantity of venison cut in long strips, ribs, etc. A smoldering fire of old stumps still sent forth its smoke upon the meat above it. Now, convinced that the owner of the place we had so unceremoniously taken possession of could not be far off, I returned to the hut where I found my *compagnons de voyage* busily engaged in preparing breakfast. I communicated to them the discovery I had made and all agreed that we had better put out immediately. Our breakfast was dispatched as speedily as possible, and once more we took up our line of march on the beach, carrying with us the remains of our late meal.

The weather was beautiful and it was long past noon ere we halted to refresh ourselves, which we did on the bank of a small bayou that empties into the Gulf about midway of the island. Here, among other things, the men produced a small calabash of whiskey which had escaped my search, and with many sly winks and jokes it was passed around, Jim O'Connor pronouncing it the only indestructible thing he ever fell in with since he joined the service.

One of the men, in rambling up the bank of the little stream, discovered some fine oysters, and on further search finding them in abundance, we kindled a fire of driftwood by means of a small pistol which we had found in the hut, together with a small quantity of powder. We made quite a feast. While engaged in this occupation, we were surprised by a couple of hunters from the east end of the island. They informed us that we had some fifteen miles to travel before reaching the place of our destination, where the *Liberty* had already arrived. According to their account, a large number of the settlers had fled to the eastern end of the island for the protection of the vessels of war, and were then encamped there.

After giving us all the information they could, we parted company and proceeded on our journey. About 10 p.m., we reached the encampment. I had forgotten to mention that when we entered Matagorda Bay, we found a number of women and children, families of men absent with the army, who had been sent on board of the brig *Durango* for safety. Among these was the wife of a sergeant named White, with whom I had become partially acquainted. Entering one of the tents, I found this lady, who invited me to make myself at home. As it was not the time for ceremony, I directed the men to shift for themselves the best way they could, and turned into my new quarters where, for the first time in a week, I succeeded in getting a night's rest.

The next day I rejoined the *Liberty* with my companions, where we had been given up for lost. After hearing my report, Captain Brown commended the course I had taken in beaching the boat, which was undoubtedly the means of saving our lives. My boat duty was at once recommenced and I was kept continually going on some errand or other. Miss Caroline had been removed to the *Durango*, there being superior accommodations for her on board of that vessel. Here she was alternately visited by the doctor and purser, much to her annoyance. Dr. Kemp of the *Liberty* was a dapper little fellow of about five feet, including his boots. Where he got his title was a subject as problematic as his knowledge of the profession. It was doubtful whether he ever pretended to either before he entered the service of the Texan government in that capacity. However, as he never ventured beyond a dose of salts, which he held to be a sort of universal panacea, in the case of our own men;

and lint and hog's lard in that of the wounded prisoners, whose sufferings were brought to a speedy termination, being pronounced incurable from the first by our worthy esculapian; he was regarded as a remarkable specimen of the faculty by the crew. Forgetting the deference and respect due to an occupant of the quarter deck, they did not fail to make Dr. Kemp the subject of many practical jokes which, owing to the absence of all discipline on board, were sure to go unpunished. This individual was more commonly called among the crew *"Multum in Parvo,"* the term by which he usually described the qualities of his favorite specific.

Unfortunately for the doctor's peace, he had fallen desperately in love with our fair passenger, whose smiles filled him with ecstasies, but whose frowns were—something else. Mr. Fischer, the purser, was the very antipodes of the fabricator of boluses—tall, lank and cadaverous, with long soap-locks hanging over his cheeks of a nut-brown hue, fine blue eyes and teeth which rivaled the snow in their whiteness. Being of a generous nature and the very essence of good humor, he was a favorite with all hands. This person was the rival, the hated rival of the former individual in the good graces of the fair lady; but although by far the most companionable of the two, he found less favor in his suit than the other. Whether she was fearful that he might make an impression or not, I could not find out. At all events, the lady treated him with incivility almost, as he expressed it, amounting to rudeness. The rebuffs which the purser met with did not mollify the wrath of the gallant doctor in the least, who had already threatened the execution of speedy vengeance upon his rival in case his visits were not discontinued.

Thus far matters had progressed when I rejoined the vessel, the lady having removed to the brig the day before, an injunction being laid upon the belligerents forbidding either to follow her. Each reproached the other as the cause of their common misfortune; and upon my reappearance, knowing that my duty would often carry me on board of the brig, my favor was courted by both parties, with the object of making me the bearer of their *billets doux*. The doctor took the precaution to sound me as to my own feelings towards his lady-love and, finding everything right, as he supposed, he made me his confidant, assuring me he was already an accepted lover. This was news to Walker and myself. By the way, the first lieutenant had seen fun ahead, and had directed me to report to him in all things pertaining to the matter.

During the day I visited the brig some half-dozen times, in the course of which I was the bearer of four notes of the tenderest description. I informed Miss Caroline of what was going on, and received permission to make whatever disposition I pleased of the missives. After she had read them, I immediately transferred them to Walker, who in his turn imparted their contents to all of the other officers, with the exception of the two heroes of the sport. The next day my agency was demanded to deliver a couple of notes. Of course I complied, and they followed the course of the others.

In the latter part of the forenoon, the doctor went on shore to visit some sick person. As he stepped from the boat, I took him by the button. "Do you know what's going on tonight, doctor?" He answered in the negative, requesting me to enlighten him. "Fischer is going to serenade Miss Caroline from the brig's boat astern, by the captain's permission," replied I.

"The devil! That won't do!"

"But that's not the best part of the joke."

"How so?"

"The lady will not be on board." The idea of this joke almost threw him into hysterics of laughter and he ran off upon his mission in high glee.

I returned on board to dinner, during which occurred the following brief conversation between Walker and the purser:

"Fischer, you are pretty good upon the flute?"

"Tolerable, but a little out of practice," replied Fischer, modestly.

"Pshaw, man," continued Walker, "you are a perfect Orpheus. You fairly make the instrument speak, if I am any judge of music. I wonder why you or the doctor haven't thought of serenading Miss Caroline. Why don't you get up something of that kind?"

"Capital idea! Excellent!" said Fischer, rubbing his hands and skipping about. "I can sit in the boat under the cabin windows—there's no order against that is there?"

"None at all," replied Walker, winking at the rest of us.

It was arranged that Mr. Fischer should be put into the boat at eight in the evening, and that the doctor should not be let into the secret. Fischer went below and practiced upon his flute all afternoon. Such strains of "Wake Dearest, Wake," as were wafted up from the cabin, would have moved the bowels of compassion in a satyr.

Just after dark, Miss Caroline, without any knowledge of what was going on, entered the boat which I had been sent with to convey her ashore. Upon landing, the commander took charge of her and, in a whisper, told me to take the doctor on board of the brig and leave him there. This was a new feature of the joke, which had not been in Walker's programme. The doctor had been drawing a bill upon the captain, so having him put on board I returned to the *Liberty*, where this new move was received with the importance due to it. While the purser was a skillful player upon the flute, his rival was no less excellent in singing. Possessing a fine voice and a nice ear for music, he had often, in company with Fischer, treated us to entertainment of high order. Possessed of this knowledge, we hoped to have a joke upon each, which should break up the folly of both the swains.

Precisely at eight o'clock, I conveyed Fischer to the boat of the brig and left him, wishing him a happy time of it. The brig was anchored but a short distance ahead of the *Liberty*, so that every movement of the parties could be easily conjectured by those in the secret. The night was chilly, but the strains of the flute sounded magnificently upon the still air and I felt already half sorry for poor Fischer, whose fingers must have suffered some. After playing some half-dozen tunes, we noticed the appearance of a light at the stern windows of the *Durango*. Again the flute sent forth its melodious strains, which, had the young lady been on board, would have moved her to pity the poor fellow who was punishing himself so much for her sake.

After an hour and a half's siege, some person from the taffrail of the vessel invited the minstrel to come on board, an invitation which needed no repetition. Fischer was introduced to the cabin of the *Durango*, where he found, instead of his fair flame, his hated rival and the mate of the brig, indulging in a glass of hot stingo over a game of all-fours. Although bursting with rage, concluding that the lady was a party to the whole affair, a belief in which he was confirmed by the doctor, he agreed to give up the chase. Upon these terms, they shook hands for the present and resumed their old friendly habits of intercourse, much to the satisfaction of the doctor who little knew that there was a rod in the pickle for him.

All further proceedings might, however, have been stopped, had it not been for the continued unmerciful quizzings of that personage, whose tongue never failed to commence wagging the moment an opportunity occurred. This annoyance towards his chopfallen rival finally determined Walker to carry out the plan originally concerted for the benefit of the now triumphant disciple of Galen.

The parties exhibited a tall specimen of polite good nature, when I re-conveyed them on board of the *Liberty* towards midnight, which continued upon the doctor's part just as long as the time occupied in transferring him from one vessel to the other.

The next day, the subject of retaliation was broached by Walker to the purser, who being necessary to furtherance of the scheme, entered into the affair with a becoming spirit. The order forbidding the doctor to visit Miss Caroline was rescinded, and he was permitted to go on board, the sanction of her father having been obtained to the part assigned to his daughter in the farce. The young lady had already been reinstated in her old quarters, where I had given her an account of what had been going on, dwelling considerably upon the chilly situation of poor Fischer on the previous night which, instead of amusing, appeared to distress her much. But when I told her of the doctor's exultation and of his annoyance to the purser, and made her acquainted with our plans to turn the tables upon him, she readily consented to become a party to the joke.

The doctor had obtained permission to give a serenade in earnest, and as the accompaniment of some musical instrument was indispensable, we all united with him in his request to Fischer to lend his assistance with his flute. For effect's sake, he at first demurred, but at length suffered his objections to be overruled and consented to play second fiddle in the enterprise. The affair was to come off that night, and everything was immediately prepared for the occasion. Fischer gloried in the possession of a pair of immense boots, such as are commonly worn by fishermen while pursuing their occupations. These reached to his hips, in which position the tops were held by small straps, depending from a belt around his waist. These he invariably kept so well sheathed with a preparation of beeswax, tallow and rosin, that, except when exposed to the hot sun, they would stand alone. The idea of water penetrating through them was preposterous. The possession and use of these appurtenances to the outer man earned for their proprietor the soubriquet of "Boots," which was, however, on account of the general esteem in which he was held, seldom applied.

The time had arrived, and the parties were assembled in the gangway of the *Liberty*, while the boat's crew were hauling the boat alongside to convey them to their destination. "What the deuce have you got those boots on for?" inquired the doctor, staring at Fischer and bursting into a loud laugh.

"Chilly work, this serenading till ten or eleven o'clock," replied he, "and then the termination," he continued in a voice which convulsed with laughter all within hearing: "O hang it, though, I'm no man to bear a grudge; here's my hand upon that." The doctor accepted the proffered grip, and we all descended into the boat. In a few moments, I saw them safely floating astern of the *Durango*, which had been stripped of oars, sails and everything else moveable.

The night was cloudy and dark, but very still, so that we could hear plainly but not see what was going on. In the course of an hour, during which time floods of music seemed to pour over the waters of the bay, the tide turned, which brought our vessel ahead of the brig. Within half an hour after this occurrence, we all dropped silently under her bow, and gained the deck just as the fun commenced. The cabin windows, darkened, were open, the boat some forty feet astern. Suddenly the music stopped in the middle of a strain of the Scotch air, "My love is like the red, red rose," and the alarmed voice of the little doctor was heard: "Where the devil is all this water coming from, Fischer?"

"From the bay, most probably," replied Fischer, coolly.

"This is some of your work, blast you!" muttered the doctor. "You've pulled the plug out."

121

"That's a mistake of yours, my friend," returned Fischer; "however, pull up and we will go aboard. If we stay here, we shall surely get a ducking."

Seizing the boat's painter, which, by the way, was carefully made fast on board of the vessel, the principal of the serenaders essayed to haul her under the stern of the brig, but to his consternation, the rope yielded and the boat commenced drifting away towards the head of the bay.

"What shall we do? Can't you find the plug, Fischer? Brig, ahoy!" yelled the doctor in one breath. "My God! I can't swim—I shall be drowned! Blast the girl, and myself too, for making such an ass of myself!"

With such ejaculations falling from his lips, he beheld with dismay the rise of the water in the boat, taking no notice of Fischer, who pretended all the time to be searching for the hole which let the water into the boat, the plug of which lay at the bottom of his coat pocket. Occasionally, he would raise himself and join the doctor in hailing the vessel, from which they were now distant the length of two lead-lines of one hundred fathoms each, which were fastened to the stern of the boat beneath the water line.

The water being now up to the seats and still rising, rendered the doctor nearly frantic with fear and rage, for the coolness with which Fischer took matters convinced him that he was the victim of a conspiracy to which Fischer was a party. The doctor had taken possession of the middle seat of the boat, consequently the lowest, and the water washed over it. He ventured to spring to the forward one which, owing to the shortness of his legs and the sudden motion of the boat (the result of a maneuver of Fischer, who watched him) caused him to miss his object, and soused him back into the water, with which the boat was now as full as Fischer thought consistent with safety. Quickly inserting the plug in the hole in the stern, which had been bored for the purpose, he managed to seize his now vanquished enemy who, having lost all presence of mind, was in imminent danger of drowning. He raised him up in the boat, where the water reached above his waist, and he stood shivering with cold, and sick from the effects of the saltwater which he had swallowed.

Considering the game nearly played out, Walker now sent me to their assistance with the boat. Just as we started, I observed another boat pulling toward them which, being nearer, reach the object of my destination first, and by the time we arrived had taken the doctor on board (Mr. Fischer declining their assistance) and pushed off. This boat was from the Commodore's vessel, on board of which the shouts of the doctor had been heard. We conveyed Fischer, now perfectly elated, on board of the *Durango*, where he received the congratulations of all hands. The commander-in-chief of the navy took the doctor under his protection, and he did not condescend to honor us with any further notice. Through his means, the whole story reached the ears of the Commodore, who was shocked at such scandalous proceedings on board of a national vessel. Entertaining no good will towards our refractory commander, he now set himself to work in earnest to rid the service of an officer who openly set his authority at defiance.

The time for which we had entered the service having long since expired, many of the men had demanded their discharge, which were not granted. So the little authority of the officers over the men had grown less and less. In addition to this, the crew indulged in a continual course of wrangling and fighting among themselves. About this time, the ship's cook, for some dereliction of duty, fell under their displeasure and, taking the law into their own hands, they proceeded to inflict summary chastisement upon the delinquent, notwithstanding the presence of the commander, who was on board. The uproar attendant upon this demonstration brought him upon deck, when, after inquiring into the matter and finding the men could not be appeased, he ordered the cook to be seized up to the rigging in true man-of-war style. He commissioned the boatswain, a burly old salt by the name of Baker, to give the culprit an earnest of the good wishes of the crew in the shape of two dozen lashes with a rope's end about half an inch in diameter.

But now came the most difficult part of the business. Neither warrant officers nor men were willing to see flogging introduced on board our vessel at this late hour. The yells of the poor fellows undergoing the punishment of the cat-o'-nine-tails on board of the *Independence* and *Brutus*, which were heard daily, came fresh to the recollection of the ship's company, and determined them to stand out against any innovation upon their already established rights.

Therefore, while the seizing of the culprit was in progress, a few words sufficed to determine them that if there was any flogging to be done under the auspices of Captain Brown, he must do it himself. Accordingly, when the boatswain was called upon to perform the duty which is assigned to him in such cases on board of armed vessels, that functionary stepped forward and, disengaging the silk cord by which the badge of office was suspended from his neck and laying it upon the breech of a gun, peremptorily refused to acknowledge the right of the commander to exact the performance of any such duty from him, alleging that he had accepted his office with the express understanding that there was to be no flogging on board of the vessel. The boatswain was a favorite with the commander, and as the latter apparently had a presentiment of the approaching termination of his naval career, he seemed to be disinclined to insist upon obedience. So directing Baker to resume his badge, he selected one of the crew, an old man-of-war's man, to perform the unpleasant duty. This man refused, on the grounds that it was no part of his duty in the first place; and in the second place, that his time having expired, he was detained on board against his will, which example was followed by all others.

The commander by this time was in a fury, and swearing that the man should be flogged, he called me as a last resource. Whether or not the men were fearful I would obey the order, there was no mistaking the threatening looks they bestowed on me, or their exultation at my following their example, which capped the climax of the captain's fury. I was the only one who had to suffer for the disobedience of the whole crew. The gunner was ordered to bring a pair of handcuffs, which being done, he repeated his order to me. I refused, and the next moment I was ironed and transferred to the sail room in the forecastle, while the commander proceeded to the undignified execution of his own order, by bestowing two dozen upon the offending cook. Shortly after, I received a call from the captain who, seeing me comfortably situated, informed me he would keep me there a week, and then left me to my own reflections.

Contrary to my expectations, I was released early in the afternoon and ordered to proceed to Point Bolivar with a boat, and procure wood for the vessel. I determined to return no more on board and accordingly made arrangements for a cruise on shore, which being accomplished without exciting suspicion, I bid goodbye to the *Liberty* for the last time.

When I returned to Galveston she had been discharged from the service and her commander was no longer recognized as an officer of the navy. There were many charges against Captain Brown, the most prominent of which (I know not with how much truth) was that we had sailed upon our previous cruise on the Mexican coast without orders or even a commission—certainly a very grave offense. This, taken in connection with our proceedings at Sisal in regard to the late schooner *Sabine*, rendered the character of our vessel equivocal, to say the least; and had we been captured by the American ships of war, our situation would have been anything but pleasant.

An hour's pull against the strong tide setting into the bay landed us at the point, where I at once made known my determination to the boat's crew, and left it to their own choice whether they would follow me, or pursue the errand on which they had come and return to the schooner. Only one, Williams, joined me, and bidding the rest good-bye, my companion and I took up our line of march along the beach, without much caring which way we wended our steps.

On to San Jacinto

Pushing along in high spirits until the sun began to get well down into the west, we mounted the sandhills and found ourselves opposite a ranch which stood about half a mile from the beach, upon the prairie towards which we shifted our course. On reaching it, we found its only tenant to be a rheumatic old sailor, a pensioner on the bounty of its former inhabitants, who had volunteered to remain in charge of the property while the proprietor sought safety beyond the Sabine River during the troubled state of the country. This proprietor was no other than the Mr. Parr I have before mentioned who, immediately upon being released from his imprisonment on board of our vessel, had moved out of reach of either friend or foe, leaving our new entertainer in charge.

The house was a shanty upon a large scale, built of rough boards, though the frame had evidently been got out by someone skilled in carpentry. It appeared dingy enough upon reaching it, though at a distance it had presented quite a respectable appearance. The interior of the house was in strict accordance with the exterior; and the posts, studding and rafters, blackened by smoke, were exposed to view. A bedstead formed of posts set in the earthen floor, upon which were fastened poles parallel to each other, across which were slats covered with a quantity of dry corn husks, spread with a large dry hide and a couple of warm blankets, presented a bed which was not to be despised by a tired man.

Our host, who welcomed us heartily and offered us the best accommodations in his power, was at this time suffering acutely from his malady, and we were therefore obliged to commence the duties of housekeeping on our own account. Therefore, while my companion prepared to display his skill in converting sundry ears of flint corn into a loaf of bread, I selected a musket from a number which stood in the corner of the house. Being furnished with the necessary ammunition by our host, I bent my steps toward a small herd of cattle which were grazing at no great distance from the house. To get within gun-shot of these animals I was obliged to exercise all my ingenuity, on account of their extreme wildness. At length succeeding, I had the good fortune to bring down a fine fat yearling heifer, from which I loaded myself with as much meat as would answer our present purposes. I returned to the house without troubling myself about the remainder, which served to regale a herd of wolves who made the night hideous with their yelping and barking.

In the morning, we bid farewell to our kind entertainer and started for the Fire Islands, some twelve miles distant, where we were told there was another ranch. As Jack Scott, our late worthy entertainer, had informed us that the muskets before-mentioned were government property, we improved the opportunity by selecting each a musket and bayonet for our defense, in case we were obliged to camp out at night. Having taken care to provide some refreshments for our journey, we did not overexert ourselves in walking, consequently it was long after dark before we reached the place of our destination. The Fire Islands, so called, we found were a couple of spots in the surrounding prairie elevated some forty or fifty feet above the level of the plain, heavily wooded, and at a distance bearing a close resemblance to green islands in a lake. Most of the beautiful prairies in Texas are dotted with similar lovely spots, serving to make the landscapes agreeable that would otherwise be monotonous in the extreme.

On reaching the ranch, we were received by the proprietor in the most friendly manner, who proffered the hospitalities of his home during such time as we wished to remain, adding that we would be obliged to depend upon his family for our entertainment, as he was about to start for Lynchburg at the mouth of Buffalo Bayou. The goods and chattels of his household were for the most part packed up, to be transferred to a number of wagons which were kept in readiness for use at a moment's warning. The removal of these, together with his family, consisting of two women and four or five little ones, he had confided to a couple of trusty slaves, themselves the fathers of a small host of woolly-heads who were peeping from every corner to see the newcomers. Finding the settler's destination was the army, I proposed to accompany him, which he at once agreed to, offering to provide me with a horse, etc. I now endeavored to get my companion Williams to join us, but to no purpose. He was determined to go nowhere but the United States. There was now nothing further to do but to provide some refreshments before starting on our journey. Having done this, we took leave of the family and, mounting our horses, my acquaintance taking the lead, we galloped off, taking a southwest direction, so as to strike the part of Galveston Bay at its intersection by Redfish Bar, some eighteen or twenty miles from the island of Galveston.

Nothing disturbed our midnight march save the howling of the wolves who, startled from their propriety, made the welkin ring as they hurried away at our approach. These animals are small, not exceeding a large dog in size, though somewhat longer in the body, and are of a dirty brownish color. In all cases where I encountered them, they proved themselves cowardly, but my conductor told me that when hungry, they were formidable antagonists. At the time of which I write, they roamed the prairies in herds of thousands, judging from the uproar they made. They are very troublesome to the settlements, especially in winter, when they steal whatever is within their reach in the shape of provisions.

About three o'clock in the morning we reached the shore of the bay, where we found a large campfire brightly burning on the sand, by the side of which were stretched the forms of three men, two of them negroes, and all buried in profound slumber. Floating on the water at a short distance was a fine large whale-boat, with its masts and sails disposed for instant use. As we approached, the voice of my companion aroused them from their sleep. "Hallo, Sykes! Hallo, boys! Come rouse up, rouse up. It's time to be off. We've no time to lose." The darkies, springing to their feet, soon stripped our horses of their furniture.

We were soon under sail for the head of the bay, where we landed at the plantation of a Dr. Richardson, where we breakfasted. We were received with the uniform kindness and hospitality which generally prevailed among the settlers and in a short time, a bountiful supply of edibles smoked upon the board, of which we were invited to partake. To me this meal was a curiosity, in which light it certainly would be regarded by northerners. Of meats there were at last a dozen dishes, including bear meat and venison, besides chicken, turkeys and ducks, with catfish from the river and red fish from the bay, and sundry dishes of hominy and rice, flanked by immense piles of corncakes and wheaten biscuits, honey, etc. The only thing I missed from the table was butter, the want of which was not felt in the present instance. Of drinks there was coffee and tea. Liquors were not produced at the table, but we had all indulged in a "smile" before being introduced to the breakfast room. The house, which was one story, was surrounded by a verandah, the front commanding a splendid view of the bay for twenty or thirty miles. Surrounding the house was a noble orchard of peach and fig trees, the staple fruits of the country. Judging from appearances, the doctor was possessed of a numerous force of field hands, some new sable-visaged personage continually passing in review before us.

About nine o'clock, bidding farewell to our hospitable entertainer, we once more started on our journey, or rather voyage, and with a fair wind we made rapid progress toward the mouth of the San Jacinto River, whose strong current we soon found ourselves breasting. The boatman Sykes pointed out to me New Washington, the settlement of Colonel Morgan, which, as far as I could see, comprised the house of that personage. About noon we came in sight of Lynchburg, of which place, from its frequent mention, I had formed a high opinion. In this I was greatly disappointed, finding nothing but one house and a sawmill on the eastern side, and the wrecks of one or two old steamboats on the other. At this place we landed, and the boat left us.

We now shouldered our horse furniture and, striking what my companion called a bee-line, we arrived at a farmhouse on the left bank of Buffalo Bayou, about five in the afternoon. Here we found no one to receive us, the proprietor and his family having taken the "Sabine shoot," as the running away of the settlers was facetiously called. There remained enough in the house, however, to enable us to make ourselves comfortable, so we concluded to remain for the night. In the meantime, my companions advised that we should provide something for our march the next day, which we did by filling a small bag with jerked beef, after roasting it.

The morning found us early on our way, and we reached the town of Harrisburg about eleven in the morning. This, the most populace place I had seen in the country, was almost deserted, a few persons still remaining at the house of a Mrs. Harris, where we were hospitably received. Everything foretold that the enemy were at no great distance;

and as the few people remaining in the place were about to leave, we took possession of an old pirogue and continued on toward the place where we were directed to look for the army. We made good headway in this sort of navigation, and toward the middle of the afternoon reached a clearing made by a Mr. Batterson. Here we remained for the night, camping in the woods as a measure of precaution.

Before quitting this place the next day, we hauled the pirogue into the bush, where she was hidden from any observation, and continued our quest on foot. Keeping along the edge of the timber, we reached the camp of General Houston about twelve o'clock, where we were warmly welcomed by all with whom we came in contact. My companion, who was very well acquainted with the General, gave me an introduction, and I had the honor of shaking hands with that illustrious personage, who requested us to attach ourselves to some corps so as to enable him to know where we were. My fellow traveler joined Captain Graham's company for the time and, my inclination leading that way, I attached myself to the artillery as a volunteer.

This part of the army was composed, including details from other companies, of about sixty men. These had under their charge two light iron field pieces (four-pounders), a gift, it was said, of the citizens of Cincinnati. Judging from the nature of the shot—old scrap iron, pieces of spikes, with a large admixture of musket and rifle balls, etc.—they were calculated at a short range to do great execution. This part of the force was under the command of Colonels Neal and Hockley. The riflemen, judging from their appearance, were men who could perform against any enemy to the satisfaction of all parties interested. Their arms, among which might be found every variety of that deadly weapon, were in perfect order. They numbered near two hundred. The cavalry were very limited in number, some sixty-five or seventy rough-looking fellows, who sat upon their horses with a sort of jaunty nonchalance which defied all discipline. The infantry were a heterogeneous mass of all nations upon earth, and I had almost said of all colors. Among them was a company of about thirty Mexicans. The head of this part of the army, the skeleton regiment called the First Regulars, Colonel Burleson presented a very soldier-like appearance; and the whole, although their equipments looked rather musty, and their clothing worse, bore the aspect of a body of men whose introduction to any field of combat would frighten, even if they could not whip, an equal number of any troops in the world.

But few members of the Provisional Government honored the army with their presence. Of these, however, the countenance and activity of Secretary of War Rusk and the dashing Lamar, who assumed the command of the cavalry corps, served to reassure the men, whose hopes of success were almost annihilated. All who were there had determined, like desperate gamblers, upon staking their last and all upon a single cast of the die, for the recovery of their waning honors.

This was the darkest hour Texas ever saw. Her army of twelve hundred men upon the Colorado had been forced to fall back, first to the Brazos, and thence to Buffalo Bayou, before the approach of upwards of seven thousand men who were advancing from three different points. Now the narrow strip between the latter stream and the Sabine was all that remained for the victorious Mexicans to wrest from the northern barbarians, when the Republic of Texas would be at an end, and the Napoleon of the south could pursue his boasted intention of planting the Mexican eagle upon the American capitol. Anxious to secure himself the honor of quashing the revolution, the arrogant Mexican pressed forward to the fatal field of San Jacinto, surrounded by his choicest battalions and scouting all idea of danger—the day being already appointed when he should receive the congratulations of his sycophants at the capital.

But I will not anticipate. It was lucky for us that we had provided for our present sustenance, as we found the army totally destitute of the means of subsistence,. My late traveling companion divided our stock of jerked meat with me when we separated, and by his advice I husbanded it with great care. We were encamped on one of the forks of the bayou, and the weather being damp and foggy, we were in a very wet and uncomfortable condition, especially during the morning. At night the want of blankets made my situation very disagreeable. But I was not alone: the shelter of each tree had its tenants, upon whose heads every sough of the wind would deluge us with a shower of water from the canopy above, where it had gathered from the heavy fogs.

The notice of the approach of the Mexicans was received on the 18th, and every preparation was made to give him a suitable reception in case he should attempt to cross that stream. Presently, however, the information was given that they had altered their course down the bayou. As soon as this was ascertained, the order was given to fall in, and in a short time the whole force was in motion toward Lynchburg. In this march, we experienced much difficulty in getting along with our artillery and baggage, on account of the softness of the soil, which was saturated with the spring rains. At those times when the wheels of the wagons mired, which they did continually, none were more willing to clap their shoulders to the spokes than old Sam himself. His "Come, boys, let's help the poor creatures," passed into a by-word among the men.

With such marching it may readily be supposed that our progress was slow, but such was not the case. No further delay was permitted than would suffice to extricate the vehicles. Towards noon, we encountered the smoke occasioned by the burning of the town of Harrisburg, fired by the order of Santa Anna. We continued our march some fourteen or fifteen miles below, where we crossed the bayou by means of a large flat, and found ourselves upon the same side with the enemy. Our course was now through the edge of the timber, and after struggling a short distance further, we encamped. The next day we moved to a point formed by a small bayou which debouches into the mouth of Buffalo Bayou called, I think, Simm's Bayou. This stream was bridged about a mile and a half from the post we now occupied. Moving to the front of the timber, we encamped within plain sight of the Mexican camp on one side and the Lynchburg ferry and shipyard on the other.

Simm's creek, or bayou, was one of those gullies formed by the draining of the land, and at the time of the spring rains it is deep and rapid. The water in it was at this time quite high. By the road through the edge of the prairie, and the bridge across the bayou, the Mexicans had gained their present position. Our camp faced obliquely that of the enemy, who as yet had no idea of our presence. This day we had been without any provisions; but after having got quietly settled in our camp, the sight of a number of beeves grazing at a short distance was too strong a temptation for our men to resist. Application was made by the men, and permission given by General Houston, to endeavor to make rations of the animals. In a short time, the prairie was dotted with the exciting chase and the sharp ring of muskets was heard in every direction, added to which was the firing in the camp. Our march the two previous days had been wet and uncomfortable, and the charges of the men were wet and useless. To remedy this, orders were issued to discharge their firearms and clean them prior to reloading. In obedience to these orders, the muskets were discharged, but for the gratification of some whim, many of the men loaded and fired their pieces a half dozen times, and the din of small arms was kept up for half an hour. This probably led the Mexicans to come much nearer the truth as to how many men there were in our camp than they would have done, and thereby prevented an earlier attack with the force Santa Anna had with him before Cos joined.

This was the first intimation the enemy had of our proximity; and supposing, as it appeared, that we were some party of Texan militia, Santa Anna ordered his cavalry to charge the wood and drive us from our hiding place, at the same time opening a furious cannonade upon that part of the prairie where our men were engaged in the pursuit of the beeves. The Mexican cavalry were handsomely met and driven back by our horsemen, under Colonel M. B. Lamar, who hurried them up until they took refuge in the rear of a body of infantry, which had advanced to cover their retreat. These, on the appearance of a part of our infantry force, withdrew within their lines, and the Mexican general contented himself with keeping up a continual cannonade upon the woods in which we were encamped, which resulted in nothing more serious than the infliction of a terrible scourging upon the trees under which we safely reposed. In this situation all were kept in constant readiness to repel any attack which might be made, but the night passed without any further alarm.

The 20th was a day of expectation, without any important results. A movement made by Colonel Sherman, evidently to precipitate a battle, was defeated by the commander-in-chief. Colonel Neal was wounded by grapeshot. Had we attacked the enemy during the forenoon of this day, we should have gained an easy victory, as the enemy's infantry had taken their arms to pieces for the purpose of cleaning them. The enemy kept up his cannonade at intervals during the day, and every preparation was made to guard against surprise during the ensuing night, the whole force reposing on their arms.

Everything remained quiet during the night, and the morning of the 21st dawned upon us—the day that was to consummate the independence of Texas. During the night, the bridge across the bayou had been cut away by a few of the scouts belonging to the army, by order of General Houston. Not, however, until Santa Anna had received a reinforcement which increased his force to almost thrice our number. Thus matters remained until between three and four o'clock, when the order was given to parade without delay. In a moment every man was in motion. The various companies fell into line and presently, emerging from the wood, directed our march to the camp of the enemy, from whom we were partially concealed by an intervening grove of timber. Making our way through the woods, our column deployed in line of battle in front of the hostile camp.

The enemy, who seemed to have been lulled to security by our previous inaction, now appeared as if determined to make up for their apathy. Upon the instant of our advance, the clang of their trumpets and rattling of their drums fell with startling effect on my ears; for this being my first battlefield, I am free to confess that I did not display the coolness of a veteran. I experienced a depression of the spirits, almost amounting to dread, from which I did not recover until the first discharge of our pieces restored me to a sense of the duty I was called upon to perform. This excitement soon left me and I saw the line was complete, and the next instant the order was given for the whole to advance.

The details of this battle have become a matter of history, and therefore I shall confine myself to what fell under my own observation. Our artillery occupied a gentle rise of ground, which enabled us to pour a murderous fire upon the enemy, who had already opened on us with both their artillery and musketry. Our line, however, continued to advance without firing a musket. It was almost wonderful to see the admirable precision with which the ragamuffins maintained their ranks during their advance upon the enemy. Old Sam had declared that he could never get his men into a straight line before he formed them that afternoon. Meanwhile, our little four-pounders vomited forth showers of destructive missiles, carrying havoc and death through the ranks of the enemy. Previous to leaving our encampment, at the end of a short address in which General Houston declared his intention of fighting the enemy, the battle cry of "Remember the Alamo!" had been given to the troops. But to return...When within seventy yards of the enemy, the line was ordered to halt and fire, and immediately after, to fire at discretion. The effect of this fusillade was terrible. The Mexicans fell like grass before the scythe of the mower, while their answering fire was comparatively harmless, their balls whistling over our heads in showers.

While this fusillade was in progress, our guns were brought into line and the infantry with leveled bayonets rushed forward upon the enemy. The charge resembled the onset of a host of demons rather than that of men. They precipitated themselves upon the appalled Mexicans with yells and shouts, like so many tigers, and above all might be heard the hoarse battle cries, "Remember the Alamo! Labadie! Goliad! Crockett! Travis! Fannin!" while in many cases the Indian war whoop was attempted with brilliant success. Before this unearthly uproar, the panic-stricken Mexicans melted away like icicles in the hot sun. The flank of the Mexican line broke and fled at the commencement of the charge, while their center made a brief stand, their position being supported in the center by a small breastwork formed of bags of sand. Behind this was planted their artillery, a fine brass nine-pounder which, at the instant of its capture, was loaded to the muzzle and was by our men turned upon the Guerreros, who still made a show of resistance.

This decided the day in our favor. The enemy were now scattered and flying in every direction. But still the work of retribution went on, and amid the groans of the wounded and dying, which were enough to appal the stoutest heart, the fierce battle cry still rang forth, and the unerring rifle and bayonet continued to swell the number of the slain. Every effort was made by Houston to stay the horrid work, but in vain; until the men, satiated with carnage, and meeting no further resistance, ceased the work of death and returned to their ranks. The whole affair had been brief, occupying but half an hour, but in this short space of time nearly seven hundred of the enemy bit the dust, and a larger number of prisoners, with their whole camp baggage and military chest containing a large amount of money, fell into our possession.

Armed with my old musket and bayonet, and acting as one of the covering party of the artillery, I had fired seven rounds from my musket and had driven home the eighth charge when I found, to my great perplexity, that I could not withdraw the ramrod from the barrel of the piece. I essayed a score of times to extricate it, but in vain. Bringing my firelock to the shoulder, I moved on with the others and got along very well, considering the circumstances, until the enemy's line was broken, an event which was quickly followed by the greater part of our own in pursuit. I started with the rest, and so closely did we hurry them along, that a party of about a dozen of the enemy, despairing of escape, turned upon us and, bringing their bayonets to the charge, advanced a couple of paces. But their courage again forsook them, and as we were about to slacken our pace to fire, they, with three exceptions, threw down their arms and fled. The three still continuing to advance, those of my companions whose guns were loaded raised their pieces to their shoulders. Without thinking of my unlucky ramrod, I did the same and fired. I did not stop to witness the effect of my discharge. It was enough for me that I saw my trusty old musket describing through the air the course of a shell just projected from a mortar, while I indulged in an evolutionary exhibition of ground and lofty tumbling, ending with a descent in the midst of a heap of dead Mexicans.

Being rather more frightened than hurt, I once more got upon my feet, and finding that the extent of my wounds was merely a lame shoulder, I determined to make a demonstration upon the enemy's camp. A large marquee close by me appeared to invite my attention, and upon entering it, the first object that met my gaze was an officer in the uniform of a colonel, stretched upon an iron camp bedstead. He was dead, but not yet cold. The appearance of things gave evidence that he had anticipated a more pleasant occupation than he had met with. It was apparent that the Mexicans, when attacked, were just preparing their evening meals, the tables for supper being left spread for the accommodation of the victors. Considering myself an irregular, and wounded at that, I did not take the trouble to return to camp until nightfall, by which time the prisoners were all gathered in, a portion of whom had to be driven from the grass by setting it on fire.

I had now no further wish to remain with the army, and anxiously awaited some means of leaving the camp. I had armed myself from the spoils of the enemy, and therefore did not return to the quarters of the artillery, but bunked in with a number of Turner's infantry. The day after the battle, I lay quiet in my new quarters and my shoulder being lame and much inflamed, I bathed it frequently with cold water. On the morning of the 23rd, taking a stroll to the old steamboat yard opposite Lynchburg, I encountered the boatman Sykes. This was the very thing I wished, and I bargained with him to be taken to Galveston, for which place we started that night.

We went a considerable distance out of our way for the purpose of landing a man and two women at Anahuac, at the mouth of the Trinity, which, together with light winds, prevented our getting to Galveston until the morning of the 26th. The steamboat *Yellowstone*, with a portion of the Mexican prisoners, together with Captain Turner's company of regulars and Graham's volunteers, had already arrived. I now found that during my absence, my old navy commander had been deprived of his rank in the service. The *Liberty* was virtually put out of commission, never having been regarded as a national vessel after the expulsion of the captain. The majority of the crew, whose time of service had expired, remained on board until she sailed for New Orleans, where she was sold out of the service.

The eastern end of Galveston island at this time presented a strong contrast to its appearance on my first visit to its shores. All was bustle and life, where formerly nothing but a dreary solitude had met the eye. The point was now whitened with the tents of the troops, whose numbers were daily augmented. Among the new accessions to the military force of the country at this time were a company of riflemen called the Buckeye Rangers from Ohio, and Bohmline's Artillery from Philadelphia. A cavalry company was also being formed under the command of Captain Stanley, who, unfortunately for the project, was killed in a duel a short time after. The particulars of this unhappy affair, of which I was partly a witness were these:

The beef for the supply of the post was procured from the northeastern side of the bay, or Point Bolivar, and on delivery, was conveyed to a shamble erected near the Commissary's quarters, where it was cut up and distributed to the various companies. The fatal quarrel originated upon a question of precedence. Captain Stanley, as an officer of cavalry (usually considered a superior arm of the military,) claimed the right of receiving the first award from the

Commissary, a right likewise claimed by Captain Graham, on the grounds of having been longer in the service, and of participating in the late actual struggle in support of the independence of the country. Neither of the party being willing to concede the point, high words followed, and the lie was given by Stanley, when, to prevent their coming to blows, they were temporarily separated and so far pacified as to restore quiet, and the quarrel was supposed to be ended. Unfortunately this was not the case, and it was not known that a challenge had been given and accepted until it was too late to stay further proceedings in the matter.

They met the next morning on the outer beach of the island, about four miles from the post, where they exchanged shots at twelve paces without effect. This failing to bring about an accommodation, Graham fired, his ball striking his antagonist just over the right eyebrow, whose pistol was discharged in the air as he fell. Thus ended this tragic affair, which deprived of the service an excellent officer, and the whole of us an esteemed comrade. Stanley was immediately brought to the camp and every effort made to save his life, but in vain. At half past four in the afternoon, he breathed his last.

The next day we buried him with military honors on the ridge of the point upon which the encampment was situated, which place became the cemetery of the post. There, too, were deposited the remains of the gallant Lieutenant Lamb, who was wounded in the Battle of San Jacinto, and lost his life through the agency of one of the infernal copper shot used by the enemy. Side by side with their conquerors lie a number of the Mexican prisoners who died of their wounds, and here repose upon equal terms in the embrace of the inevitable and final conqueror, death. The great gale of 1837 drove the waters of the Gulf with such force over this point of the island as to completely change its features and obliterate every trace of the camp burying ground, together with the immense heaps of mud and clay, which the Mexican prisoners were employed in throwing up, and which were dignified by the title of "Fort."

The post was at this time under the command of Colonel Morgan, who likewise officiated as a civil magistrate. Besides the garrison, there was a large number of families who, as soon as the news of the retreat of the remainder of the Mexican army to the Rio Grande reached the island, returned to their homes. Among these were many of my acquaintances, with whom I was very loath to part. Above all, I was obliged to decline the kind invitation I received from Mr. C., who wished me to accompany his family to their home upon the Colorado and become a member of their family. We parted with mutual expressions of regret, but not until he had exacted a promise that I should visit them as soon as possible, which I was unfortunately never able to fulfill.

The place was soon deserted by all but the garrison, which in its turn was reduced to two companies—one of the regulars and one of the volunteer artillery, and the routine of military life began to grow insipid. The great question with me now was whether I should still remain in the service. I had but little inclination to serve under Commodore Hawkins or Captain Hurd, and therefore returned to Lynch's and was accepted for service, receiving permission to attach myself to the *Invincible*. As that vessel was not in port, I returned to Galveston by the steamer and reported myself to Commodore Hawkins on board of the *Independence*. The Commodore questioned me pretty sharply as to the course I had pursued since leaving the *Liberty*, and advised me to remain on board until he could receive instructions from Mr. Fisher, the Secretary of the Navy, in what capacity to employ me.

Not liking this arrangement and knowing that he entertained no good will towards the refractory crew of the

Liberty, from whom he had failed to procure a single volunteer, I left the vessel without any oppositions and took up my quarters with the adjutant of the port, Mr. Hunter, who kindly gave me an invitation to that effect. My means of living had happily kept pace with the opportunities I enjoyed of adding to them, and anyone able to supply the necessary funds for the procurement of the "ardent" was a welcome visitor at Galveston.

I passed some two months very pleasantly in hunting and fishing, and always found a large supply of old Monongahela at the ferry house, situated at the west end of the island. At home, the usual amusement was cards, which most commonly ended in drunken revel. This occurring at night, of course the knowledge never came to the ears of the commandant. The fishing around the shores of the island was glorious. Mullet swarmed upon the shoals in countless millions and, with the assistance of a cast-net, we frequently caught half a bushel at a throw. Angling for red fish was likewise a favorite amusement, and it nearly cost me my life one time, the circumstances of which I will here relate, first giving the reader an idea of the *modus operandi* usually observed:

First, a line is procured varying, at the pleasure of the sportsman, from thirty to forty fathoms in length, to which is attached a single hook, somewhat stouter than the larger size of cod-hooks. A noose at the end of this line is then slipped over the hand, and a fresh mullet of suitable sizing being impaled on the barbed instrument, the operator usually wades into the water, if the shores are flat, until, as at our best fishing-ground, the water reached to the middle thigh. Then the line was cast as far in advance as the skill of the sportsman could throw it, and it was seldom the case that more than five minutes elapsed before a victim was hooked, when all the fisherman had to do was to turn and make his way to the shore, dragging his prize after him. The best season for this sport is from the middle of spring to that of the fall. But on to my story...

Arming myself with the line, bait, etc., I proceeded to the outer and most eastern point of the island—a place, by the way, I had never visited before and which, from its sandy appearance, I concluded might afford good sport. Having reached the spot, I found, to my great disappointment, that the place was unfit for fishing, as at the distance of forty feet from the shore there was very deep water. Determined to try my luck, I waded in to the depth of my knees and threw out my line. I had not long to wait. Something of a dead, heavy weight seemed to have taken hold of the hook, and thinking all was right from its partial yielding to my pull, which enabled me to reach the watermark of the shore, I began to anticipate a prize. But at this point, the object stopped suddenly and began to draw slowly off again.

Wondering what it could be, I continued to give it line, of which I had considerable slack in my hand, occasionally gently trying to check its further escape from me, which I found could not be done. I had but one coil left upon my wrist, when I realized that I was nearly to my waist in the water, and that I could not swim, and the line was fast to my wrist by a slipnoose which was drawn tight already. I got my knife in my hand in an instant, and it was none too soon. The strain drew the line tighter and tighter. I was up to my armpits and it would not do to go further.

I cut the line, and the next instant an enormous porpoise rose to the surface at about the length of my line from me and, turning flukes, he disappeared, carrying my fishing apparatus with him. I was so astonished that I could not make out, as his ugly length passed in review before me, whether he had my hook in his mouth or whether, as was probably the case, it had caught in some part of his shining black carcass. These fish, the shape of whose nose is like that of a hog, are known to resemble that animal in their

145

great propensity for rooting, in which way they principally gather their subsistence from the worms which abound in the sand and mud at the bottom of the sea. The individual who had so recently endeavored to introduce me to the mysteries of the Gulf was probably engaged in the agreeable occupation of procuring his breakfast when he either swallowed my bait, or got it hooked into his covering of fat or blubber (another peculiarity in which they resemble a porker) and thus effectually turned the tables upon me. I need hardly say I did not venture upon this dangerous fishing-ground again.

While the season lasted, our table was bountifully supplied. Oysters, with which the bay and bayous of the island abounded, were a common luxury. Shrimp of a large size and excellent flavor were drawn from the waters of the Gulf by the large seine belonging to the garrison, accompanied by great numbers of speckled trout and large crabs. Beef, bacon and venison, with snipe and curlew, afforded a bill of fare which might appease the appetite of an epicure. As the season wore on, and the season of northers set in, the various lagoons, bayous and parts of the bay, literally swarmed with geese, ducks and the beautiful swan, which winter upon these shores, insuring rare sport and a loaded table to those who were not too indolent to avail themselves of the opportunity to replenish their larders. Nothing suited me better and most of my time, for some three months, were devoted to these occupations.

No More to Do with Texas

At length the order came which was to break up this round of amusement, and gave me a chance to earn the rations of the government, upon which I had to a small extent been drawing. This order directed me to report for duty on board of the *Brutus*, then lying in port. Although not well pleased with the arrangement, I obeyed, and soon found the rigid discipline of a man-of-war a different affair than I had anticipated. The free use of the cat for flogging at first disgusted and shocked me, but this wore off and, as our time was mostly occupied in running from one harbor to another, the summer and fall passed away almost without notice.

About this time we sustained the loss of the schooner *Independence*, formerly the flagship of Commodore Hawkins, who, I had forgotten to mentioned, died in July of this year, 1837. This vessel, which was a very dull sailer, had left New Orleans for the Brazos, having on board William A. Wharton, Commissioner of the Texan government to the United States government. Captain Wheelwright, whom I have formerly mentioned in connection with the *Liberty*, had assumed the command of the vessel upon the decease of the commander-in-chief and, under instructions from the Commissioner, had proceeded on the voyage. Unfortunately for them, they encountered two Mexican men-of-war brigs when almost within sight of their destined port. The dull sailing of the vessel, together with the fact of being short of hands, and the immense superiority of the enemy's force, all conduced to render escape hopeless. After a running cannonade of short dura-

147

tion, she was obliged to strike her flag. In the affair, the commander was severely wounded.

The firing was distinctly heard at Galveston, and shortly after we received orders to put to sea and bring the enemy to action. Taking on board a number of troops, who volunteered to act as marines on the occasion, we made sail, accompanied by the *Tom Toby* privateer, under the command of Captain Hoyt. But it was beating the bush after the bird had flown. After cruising about for a day and night in the latitude of the Brazos, south of which we did not venture for reasons best known to the commander, the morning of the second day found us becalmed and surrounded by a dense fog, which continued till near night. Towards sundown a brief cannonade was heard to the southward, in which direction, as soon as the wind sprang up, we made sail. Being but a short distance off shore, we made the land in an hour's time, at the mouth of the Brazos and, running down, ascertained that the firing had been the salute of an English war vessel, which had arrived shortly before. The next day we returned to the harbor of Galveston, and thus ended the heroic search for the enemy, the counterpart of which may be found in the fact recorded by the poet that, "The King of France, with ten thousand men, marched up the hill and then marched down again."

There was neither honor or profit in remaining on board of the vessel, and I petitioned for leave of absence for a short time. That being granted, I received permission from the Secretary of the Navy to attach myself to either of the vessels belonging to the service. While hesitating which to choose, the *Brutus* sailed on a cruise and I again visited Galveston, and from thence proceeded to Velasco at the mouth of the Brazos, where I endeavored to obtain a situation on board of the *Invincible*. In this I failed, that vessel having, as was always the case, a full complement of men.

I therefore returned to Galveston, and remained with the garrison the most part of the winter, performing military duty.

Spring came, and as the news of the general furlough granted to the army had reached the island, the men were clamorous that the same indulgence should be extended to them, and addressed a petition to the government to be allowed their rights. Their prayer was disregarded, and a new commandant ordered to take charge of the post. This officer was Captain Turner, now promoted to the rank of lieutenant-colonel, a man who had won the respect of everyone under his command. Under the new officer the men submitted for awhile.

This calm was at length broken by the introduction into the camp of a barrel of whiskey. How it came there very few could tell; but in the morning, when the drums beat the reveille and the men turned out to answer the roll call, they discovered the barrel of whiskey in the center of the cantonment of the company of regulars, and instead of putting it under guard as good soldiers should have done, they commenced putting it under their jackets. This was a direct infringement of camp regulations, and was followed by the cessation of all military duty for a couple of days, during which time the whole camp, in many cases including officers, was one scene of drunken revelry. The liquor being used up, order was again restored, and the men returned to their duty.

A court of inquiry was immediately organized, to account for the sudden appearance of so unusual a phenomenon and, if possible, to detect the author of the mischief. None but good men and true were in the secret, and the court, after a session of three days, rose and gave it as their opinion that a barrel of whiskey had been introduced within the lines of the camp; but how it came there and where the devil it had gone was a subject equally mysterious to them;

but that the probability was that the source of mischief was by this time used up, and any further investigation therefore useless. With this verdict, the whole matter was dropped.

This glaring outrage upon military discipline had the effect to open the eyes of the commandant to the danger he incurred by resorting to coercive measures to restrain the men who were, as he discovered, determined to resist any infringement of their rights by the government. Two-thirds of the men in the army who had received an unlimited furlough were volunteers, six-months men, or who had not been much longer in the service; while the company of regulars in garrison on the island had been in service nearly two years, in some instances suffering great privations from lack of subsistence and exposure to the weather, yet bearing all with the utmost cheerfulness.

It was the duty of the government to relieve this company by sending a corps of the short-term men to the island. The artillery company at the post being volunteers, and not having been so long in the service, could have no reasonable objection to remaining longer in the garrison. During the riot, the men had expressed their determination to the colonel to proceed to Houston, and there make a demand for their furlough from the president in person. Fearing they would put their threat into execution, the commandant compromised the matter with them by promising that, if they would wait three days longer and perform their duty, at the expiration of that time, if no relief should arrive, he would assume the responsibility of granting three days' furlough, to enable them to proceed to Houston to make their demand. This was satisfactory to the men, and for three days the duties of the garrison were performed in a manner worthy of its palmiest days.

No relief arrived, and the time having expired, the company, after turning over their arms to the commissary, left the post in a body and proceeded to the city of Houston, on boats provided for the purpose. The city of Houston was at this time just springing into existence. Some twenty houses were nearly finished, including the capitol and president's house, which latter, by the way, was rather a small affair. Where, but a couple of months before, nothing but the long grass and primeval forest met the eye, large stores and dwellings seemed to have arisen as if by a wave of the magic wand of an enchanter; while the sound of the hammer and saw echoed strangely through the still depths of the wooded banks of the bayou, where formerly naught had broken the solitude save the bark of the wolf or the wild yell of the panther. The gambling houses and drinking saloons were already in full blast, from which the sounds of uproar and revelry nightly extended into the "wee sma hours ayent the twal;" presenting a queer contrast to the yelping devils, the wolves, who ventured to the outskirts of the town and made the air resound with their odious din.

The morning after their arrival, the company marched in a body to the presidential mansion and, forming before his door, were there met by the chief magistrate of the republic, who demanded who they were and what they wanted. On receiving an explanation of the wishes or rather demands of the men, the president very coolly told them they were sent to Galveston to support and defend that post, and advised them to return and resume their duty. He swore that in case of non-compliance, he would have every mother's son of them hung. This was sufficiently plain for all to understand, and they withdrew and returned to the island, where they remained some four months longer before being relieved.

After remaining in Houston two or three days, I took a trip to Anahuac, and from thence to Liberty, upon the Trinity. During this excursion, having exposed myself much to the hot sun and chilly night air, I contracted the fever and ague, which is very prevalent on the low levels near the Gulf. From this disease I suffered greatly for the space of three months, experiencing its attacks in all their various features, until I was nearly reduced to a skeleton. With me it appeared to defy the usual remedies. I took calomel and quinine and castor oil in large quantities, without obtaining any relief. My nervous system was reduced so low, that I experienced continually all the horrors of delirium tremens. When hope had nearly left me, a friend brought me a bottle of Rowand's tonic mixture, the first dose of which broke the disease, and in a short time I recovered my health and strength again. By the time I was sufficiently strong to leave the house, the Board of Land Commissioners were in session for adjudication of claims, and I immediately presented my land scrip, proved the date of my entering the service and received my warrant to locate.

Having but little inclination to turn farmer, I associated with a person by the name of Hoffman, and commenced keeping a sort of camp restaurant in the woods on the bank of the bayou. This proved a very lucrative business, having frequently from seventy-five to one hundred boarders at a time. The only drawback was that we were obliged to take the government paper, which was worth but five cents per dollar, in payment. I had been collecting this paper ever since its first emission, as it bore twelve per cent interest per annum until redeemed. At the expiration of five months, I was possessed of a large amount. It was with but little satisfaction, therefore, that I learned that Congress had passed a law which prevented any scrip from being audited unless the persons presented it had received it directly from the government. All I had of this description did not amount to one hundred dollars, while I had thousands that I had received from others, for goods which I had paid for in hard coin. But there was no help for it, and I lost no time in selling out my interest in the Goose & Gridiron Retreat, the name of our establishment. After disposing of my land scrip at the usual rates given for those documents, I packed up all my effects and, proceeding to Galveston, took passage on the schooner *Virgil*, for New Orleans.

Upon arriving, I at once endeavored to get rid of the government paper, and finally parted with it at the rate of two and a half cents per dollar. I was possessed at this time of nearly three thousand dollars, and I determined to have no more to do with Texas.

SAM HOUSTON VS.
S. RHOADS FISHER

The foregoing cruise reports in this chapter relating to the 1837 Yucatan Campaign aren't just about officers, ships, captures and flag-planting. They have an added ingredient: the boss is on board one of the vessels.

After witnessing the capture of the *Independence* along with her passengers and crew at Velasco, and seeing the peoples' response to that event, Secretary of Navy, Samuel Rhoads Fisher, opted to accompany the *Invincible* and *Brutus* on their upcoming cruise to the Mexican coast. He was there in the capacity of a passenger and, when needed, as a volunteer. It was his hope that the President and Congress, seeing a fellow bureaucrat take on the perils of an ocean cruise, might change their views toward the dwindling, underfunded and stymied navy. He was wrong.

Furthermore, President Houston had opposed any direct action to attempt to free the *Independence* prisoners but Fisher and the Congress disagreed. The Secretary departed with the remains of the Texas fleet on June 10, 1837, reporting back for his cabinet duty on September 4. President Houston admonished him before the Senate and removed him from office. The Senate disagreed, on October 11, calling the move "disrespectful, dictatorial and evincive." They reinstated Fisher a week later and he, in turn, informed the acting Secretary that he was back in business. Acting Secretary William Shepherd begged to differ and refused to turn over the papers of the department without Houston's approval. Fisher reported this to the Senate.

On October 26, Houston announced that he was preparing charges against Secretary Fisher and these were evidently read before the Senate on November 7 but never made it into the official records of that body. Fisher's trial before the Senate ran from November 23-27. William Fairfax Gray and David S. Kaufman represented the prosecution, while Fisher was defended by David G. Burnet and John A. Wharton. Colonel Wharton delivered a gritty closing argument that hinted at a secret thwarted impeachment attempt of President Houston and let loose a torrent of indictments upon Houston's character. It must have been a hair-raising scene to have witnessed.

At the end of it all, Fisher was acquitted of all charges, the Senate proclaiming that President Houston did not present sufficient evidence to prove he was guilty of dishonorable conduct. Fisher's removal from office, however, was upheld on the grounds that His Excellency needed to be able to appoint a cabinet member with whom he could "cordially unite."

David S. Kaufman, a young lawyer fairly new to Texas at this time, and John A. Wharton, the "keenest blade of San Jacinto," both delivered their closing arguments on November 25, 1837. Kaufman, while en route to Nacogdoches from Houston, submitted his speech for publication, likely with the *Texas Chronicle* of Nacogdoches; it is not to be found in the *Telegraph*. When Fisher learned of its publication, he requested that Col. Wharton respond in kind. The *Telegraph* had not the space to publish Wharton's oratory all at once, so they instead issued it in pamphlet form in the spring of 1838. It appears here for the first time in print in its entirety since that date.

Secretary Fisher had been a part of the final voyage of the first Texian navy and his career in conjunction with that department ended in a shoot-out with Sam Houston. Were it not for his death in 1839, one might think he and Commodore Edwin Ward Moore might have been fast friends as they had much to talk about in this regard.

SPEECH

OF THE

HON. JOHN A. WHARTON,

IN DEFENCE OF THE

HON. S. RHOADS FISHER

DELIVERED BEFORE

THE TEXIAN SENATE,

NOVEMBER 25, 1837

CORRESPONDENCE

Matagorda, April, 1838

Col. John A. Wharton:—

DEAR SIR—Not having felt disposed to take any step which might have the least appearance of obtruding my affairs upon the public attention, I declined, in opposition to the opinion of many of my friends, to make a call upon you for a copy of your speech as my counsel during the prosecution of President Sam. Houston against me, as Secretary of the Navy; but having perceived in the public prints a *speech* said to have been delivered before the Hon. the Senate, by a *young gentleman* of the name of Kaufman, who appears to have been in labor with it during his ride from Houston to Nacogdoches, and was delivered of it at the printing press, I have concluded it would be correct that the public should have the opportunity of seeing the picture as well as the *canvass*: you will therefore confer upon me a particular favor by forwarding your argument upon the subject to the Editor of the *Telegraph* for publication. I am aware of his reluctance to make controversial publications in his paper, but I am also aware of his candor and high sense of justice, and do not doubt he will, with great pleasure, adorn his columns with its argument, its *truth*, and its classic beauty.

I have the honor, &c. &c,

S. RHOADS FISHER

Brazoria, April 5, 1838

Hon. S. Rhoads Fisher:—

DEAR SIR—I have received today your letter requesting a copy of the speech which I delivered before the Senate; it was not my intention to have published it, but inasmuch as Mr. Kaufman has published the speech which he made against you, I deem it but an act of justice to afford you an opportunity of adopting the same course. I regret that my time and circumstances did not permit me to give that attention to the subject, which merits it required. I shall, however, have gained the only reward at which I aimed, should I succeed in placing your official conduct in a proper light before an impartial public.

Yours truly,
JOHN A. WHARTON

[Having received this speech during the session of congress, we should have been unable to insert it in the *Telegraph* unless in several succeeding numbers, by detached fragments; we have therefore preferred publishing it in pamphlet form.—*Editor of the Telegraph*]

Sam Houston vs S. Rhoads Fisher

Upon me devolves the task of closing the defence, a task both pleasing and painful—pleasing because it affords me an opportunity of standing up in the defence of the rights and liberties of a citizen, when threatened by all the weight and influence of Executive power and patronage: painful because it compels me to enter into disgusting details, to expose and comment upon the conduct of our Chief Magistrate; conduct at the very mention of which every patriot citizen must hang his head in shame and sorrow.

Mr. President, this is nothing more nor less than an impeachment of the Secretary of the Navy, and we are now trying him for high crimes and misdemeanors, on charges preferred by the President. Disguise it as you will, twist and torture it as you may, "to this complexion must it come at last:"—had it pleased his Excellency to have requested the Secretary of the Navy to resign for the sake of harmony in his cabinet, or for any cause except malfeasance in office, without threats or charges, I should have been saved the necessity of appearing before you today, and our country the sin, the shame and the sorrow of the transaction now passing before us.

I assure you that no desire to retain the office impels the secretary to appear before you; but merely to defend his reputation, which is dearer to him than all the world; besides the President has so insidiously interwoven his threats and his charges with his request for the Secretary of the Navy to resign, that the Secretary could not abandon his office without admitting the truth of the charges. I trust every one of you will view this matter in its proper light, and see the object for which we are contending. It is reputation, not office.

161

Fortunately for us, however, the Secretary of the Treasury testifies that it was the intention of the Secretary of the Navy as well as himself, to have resigned so long ago as last spring; and that nothing induced them to remain in office but a desire to sustain the President, he being threatened with an impeachment. This evidence, coming from so high and respectable a source, must be conclusive on this point. And after all, what has been the reward of the secretary, for his disinterested and patriotic adherence?—the abuse, the threats that are to be found in this evidence, the charges contained in this message.

However reckless may have been the career of our accuser from his cradle upwards; however notorious he may have become for calumny and detraction; yet, holding as he does the high office of the president of this country, it imposes upon me the imperious duty of treating with consideration every charge he has made, however false and frivolous it may be. These charges have gone abroad; they are already coextensive with the limits of our Republic, and doubtless, will prove coexistent with the duration of our government. These charges strike directly at the honor and reputation of the Secretary; and your decision upon them either attaches infamy to him, or brands CALUMNIATOR upon the forehead of his accuser.

Mr. President and gentlemen of the senate:—It will be my object to take up these charges in the same order in which they are to be found in the message [of the President], and to apply the evidence. I trust that I shall be able to establish to the satisfaction of each and all of you that some of these charges are absolutely false, some both frivolous and contemptible, and others, if true, are highly creditable to the Secretary of the Navy. I trust that I shall be able to establish to the satisfaction of each and all of you, that these spring from a foul and polluted source; that they emanate from a man who has long since mani-

162

fested an entire extinction of all moral principle; a man who has shown on this, and other occasions, an "inveterate blackness of heart, died in the grain with malice, vitiated, corrupted, gangrene to the very core."

The President in his message of Nov. 2, which contains the charges, after accounting for his delay, says, "However reluctant I may feel in adopting the requisite measures to sustain the character of the country, and to avoid the imputations that might be hurled at it from abroad as well as at home, I nevertheless feel constrained to submit the facts and circumstances connected with this case, without the slightest coloring." And he refers your Honorable body "to the documents accompanying this communication" in corroboration of his *uncolored facts*. Now I am not more surprised at his modesty in submitting facts without the slightest coloring than at his hardihood in boldly avowing his reluctance to adopt the requisite measures to sustain the character of the country.

However, these are small matters, and I shall proceed to state the first charge, which is in these words: "I have derived from undoubted authority that shortly after the constitutional organization of the constitutional cabinet, that the Hon. S. Rhoads Fisher, then, one of its members, and having the direction of an important arm of national defence, the Navy, addressed to Thomas Toby, Merchant, of New Orleans, a letter, desiring him to purchase tobacco, for the purpose of smuggling it into the enemy's ports; suggesting the advantages of such a commerce: and furthermore, that he as Secretary of the Navy, would participate with Mr. Toby in the transaction, and that the Navy of Texas should be so disposed of as to give protection to the traffic."

After reading this charge, I examined the accompanying documents; but not finding one jot or tittle of evidence to corroborate it, I thought the President had well said that he "submitted facts without the slightest coloring." Yes, coloring of truth! The only evidence which we have, or can be had, to substantiate the charge, has been voluntarily furnished by the accused! It consists in a letter to Thomas Toby; and from an attentive examination of this document, I believe every man of understanding will agree with me that never was the meaning and intention of any one attempted to be so misrepresented or perverted.

In the first place, the President says "he wishes Mr. Toby to purchase tobacco for the purpose of smuggling it into the enemy's ports." The idea of a Texian citizen smuggling into the enemy's ports is preposterous. No one that knows the Mexican character could ever tolerate the idea of either trusting himself or property to their mercy or sense of justice. Besides, for the honor of my country, I do hope that the President cares but little, however numerous may be the infractions of the revenue laws of Mexico, or however great the impositions practised upon her custom houses, so that in these infractions and impositions no arms, ammunition or munitions of war, no articles of clothing or provisions are introduced into the country: for in such cases Mexico alone would be the sufferer. She would thereby be deprived of a portion of her just revenue, and would in no shape or manner be better calculated to sustain herself, or carry on the war.

But suppose the President meant smuggling into *our* ports. To this I would reply, that at the time of the proposed introduction of the tobacco, we had no tariff levying impost duties, consequently there could be no smuggling into our ports: and I am forced to conclude that the word smuggling was introduced for the purpose of giving a *false coloring*, and casting odium around the transaction.

The President next says the accused prepared to participate in the transaction with Mr. Toby, in his capacity as "Secretary of the Navy." To this I reply that it is absolutely false; that he never did propose as "Secretary of the Navy:" he proposed in his individual capacity. And even to this, I am aware that there are some who would make objections, on the ground that one so high in office should not embark on commerce: but when we take into consideration what some sections of our country have suffered from the calamities of war, how much the war has impoverished our citizens—and amongst the number the Secretary of the Navy was a sufferer; and when you further consider that his salary was merely nominal, it being paid in government paper, which at that time was almost valueless; and further consider that he had a family to maintain, I hope that even the most fastidious will not object to any officer of our government embarking in honest traffic, provided it does not call too much of his time and attention from the business of his office, nor work to the injury of the country.

The President next alleges that the Secretary promised that "the Navy should be so disposed of as to give protection to the traffic." This is another abominable falsehood. The Secretary never did promise to dispose of the Navy; he merely stated a fact, namely, that a government vessel would be on the coast, watching the movements of the enemy. In this instance, let us charitably presume that the President has only mistaken the effect for the cause. And here, gentlemen, let me ask with all due deference to the opinion of our President, where should our national vessels be except on a cruise on the coast, watching the movements of the enemy, and protecting our commerce? But that you may better understand this subject, I will read you the entire letter containing this most abominable proposition.

Columbia, January 9th, 1837
Thomas Toby, Esq., New Orleans

MY DEAR SIR:—I feel disposed to make a proposition to you which may result in a very handsome profit. Tobacco is now worth on the Rio Grande $150 per bale, and if you could send out 500 bales on our joint account, I would be glad to join you in the speculation. I shall have a government vessel on the coast, watching the movements of the enemy, so that the tobacco vessel will have nothing to fear from Mexican cruisers; we can sell considerable for cash, but can trade a good deal for horses and mules: these we can sell to the government at a handsome price, who are much in need of them, and cannot get them for their paper, unless it be from some citizen who not only wishes well to his country, but is willing to put confidence in them. Suppose we sell 350 bales for cash, even at one hundred dollars, it will be $35,000; and that we trade 150 for horses and mules at $20; that would be 750 horses, which the government would gladly take at $20 each, say $22,500, making $57, 500. I believe a very extensive business could be done; and if you feel disposed to enter into it, write me immediately, and I will make the necessary arrangements. There are some Mexican prisoners here recently from the Rio Grande, from whom I get the information.

I remain very truly, &c.,
S. RHOADS FISHER.

P. S. Our planters would be glad to take our mules, if we have an excess, at $100 in government paper.

Now, gentlemen, after having probed this matter to the very bottom, let me ask you where is the guilt, when is the crime committed, or even contemplated? Does it consist in the fact that an officer of our government proposed to embark in commerce? Does it consist in the fact of the Secretary so disposing of the navy as to protect our commerce and watch the movements of the enemy? Is it to be found in his offering to sell to the government horses, of which, we stood in so much need, for one-fifth part of the amount he could have readily obtained from individuals? Again, and again, I ask where is the crime?

I have probed this matter to the very bottom; I have drained it to the very dregs; I have sifted it to the last sand; and with the aid of all the lights and knowledge I have been able to bring to bear on the subject, I have entirely failed to discover the least particle of guilt, either meditated or performed.

But I have not yet done with this subject: I am aware that the very fact of trading with the enemy implies guilt. Let us see how far this particular traffic was considered criminal, or even injurious to our interest in the estimation of some of the highest officers of our government. General Rusk states that when he commanded the army, he entered into similar arrangements with the enemy from public considerations, from motives of policy alone. Colonel Cazenau informs you in his evidence, to which I invite your most particular attention, that when General Felix Huston commanded the army, a similar arrangement was then contemplated; that the president actually issued verbal orders to him to assist Colonel Seguin *in that business* in *protecting that traffic*. If, then, these high functionaries could embark in or countenance that traffic with the enemy without crime or guilt, might not the Secretary of the Navy presume he could do the same? But, understand me, I do not attach guilt to any of the parties alluded to; on the contrary, I believe that the measure was justified by the soundest policy. We were furnishing the enemy with an article which for all the *useless* purposes of life was of the most worthless character; and we were to receive in return horses, of which we then stood so much in need: for at the time of the proposed arrangement, our army, either for the purpose of an advance or retreat was destitute of the requisite number of horses to transport the sick, artillery and baggage wagons. Superadded to all, the openness of the transaction precludes all idea of guilt. The proposition was made to an agent of our government, and the whole matter had to pass in the view of all our army, consisting of more than 2,300 men under arms.

I have now, gentlemen, in a cursory manner alluded to all the points that have occurred to me as connected with the first charge. Doubtless, some have escaped my attention, for I was engaged in this matter at so late a stage of the proceedings, that I had scarcely time to examine the evidence. Should there be any one point upon which you entertain the slightest doubt, I beg, I conjure you to name it whilst I occupy the floor; for humble as my pretensions are, and limited as are my abilities, I make bold to assert that I am every way competent to establish the innocence of the accused, beyond even the possibility of a doubt. After all, let whoever may try to cover this matter with falsehood; let the President endeavor to pervert it as much as he pleases, it amounts only to this—it was a transaction only contemplated, into which neither guilt or crime entered, and of which the leading inducements were private gain and public benefit.

I now proceed to state the second charge, which even in the manner the President represents it, does not amount to embezzlement of the public funds; nor does it show a want of integrity on the part of the Secretary. I will read the charge verbatim:

"Another circumstance, which in the estimation of the Executive, demands the attention of your Honorable body, is the fact, that upon application of the Secretary of the Navy, the steamboat *Cayuga* and the *Pocket* (a brig taken at sea,) both lying in Galveston Bay, were ordered to be sold, as it was impossible for the government, without means, to repair either one of these vessels. They were daily incurring expenses and going to decay. In part payment for the sale of these vessels, the Honorable S. Rhoads Fisher, who at the head of the Navy Department had been entrusted as the proper officer to make sale of them in the most favorable manner for the benefit of the Treasury, actually received in cash $1,600, and has never accounted for it to the government, though twice called upon by the Treasurer, except $500 which was appropriated for the purchase of provisions for the army, retaining in his possession $1,100, and that at a time where there was not one dollar in the Treasury."

By an examination of the account of the Secretary of the Navy with the government, which has been *approved*, you will find that not half the sum alleged by the President remains in his hands; and even this pitiful sum he has on different occasions, and to different officers of the government offered to pay over. The Treasurer states in his evidence that the Secretary of the Navy, on leaving Houston, made arrangements with Colonel Lewis to pay the amount due, and that Col. Lewis promised to settle the amount with him, and on the return of the Secretary, finding that it was not paid, he called on him and desired a settlement, which the Treasurer declined, on the grounds that his papers were with Mr. Welshmeyer, and could not be had. The Treasurer further states that the Secretary offered to pay him in Treasury notes, which he declined receiving, notwithstanding there was an express law declaring Treasury notes should be receivable for all government dues. The Secretary of the Treasury in his evidence, states that the Secretary of the Navy, soon after receiving the money, called upon him and offered to him the amount he had received for the *Cayuga* and *Pocket* in bank paper, and that he declined receiving the same; that he advised him to let some of the clerks and heads of the different departments have some of the money, because, says he, "we had either to furnish our clerks with means of subsistence, or go without them."

Now, if you will but take the pains to examine the account of the Secretary of the Navy with the government, and which has been approved, you will find that the Secretary did pay to the clerks and heads of the different departments more than two-thirds of the amount he received; and that the immaculate President himself received $300 of this money. And now let me ask, when, in all the proceedings, can you discover any criminality on the part of the Secretary of the Navy? Does it consist in offering Treasury notes to the Treasurer?—or in offering the amount all in bank paper to the Secretary of the Treasury? Or does it consist in advancing some of the money to the clerks of the different departments, those faithful public servants, who actually stood in need of it for the purpose of procuring their daily subsistence? Surely, surely, you will say there is no crime in all this.

Gentlemen, were I in a court of justice, making the defence of the accused, I would not condescend to reply to this charge, nor would I do it now but it proceeds from a man, so high in office and is gravely preferred to this dignified body for the purpose of removing from office one of our highest functionaries, and of destroying his reputation. Can anyone, however dull of comprehension he may be, entertain the idea for a moment, that it was the intention of the Secretary to embezzle the public funds or defraud the Treasury? If there be such a man, to him I say bear in mind that the government was indebted to the Secretary for money advanced, and for services rendered, even allowing that he possessed the *quo animo*, how on earth could he effect his purpose? My curiosity is excited to learn the *modus operandi*. I have often heard of debtors defrauding their creditors, but never before of a creditor defrauding his debtor.

Gentlemen, when I consider from what source this charge emanates, I estimate it as a high compliment; for it is tantamount to an admission that even in the dark recesses of a calumniator's heart, nothing is found greater than this: and 'tis evidence that the fountains of pollution are almost exhausted, even to their sources.

Having now disposed of the second charge, it only remains for me to examine the third and last, with the various specifications contained in it. This charge amounts simply to this:—that the Secretary of the Navy took command of our squadron, and was absent on a cruise for several weeks; during which short time our little navy harassed the enemy more, destroyed more of their property, captured more prizes, and actually performed more than had been accomplished since the commencement of the war by this arm of the national defence.

Really, if it was not for a vein of malice and envy tinctured with falsehood discoverable in this part of the message, a plain man would believe that his Excellency was setting forth the conduct of the Secretary of the Navy not for your censure but for your commendation. He tells us of the exploits of our little navy; he tells us that the thunders of our cannon were heard seven hundred miles beyond the Texian main; would to God that they could be heard in every sea! He tells us of islands being taken—officers created—oaths of allegiance administered—the national salute fired—the Texian flag hoisted—enemy's vessels destroyed—towns attacked—contributions demanded—prizes taken and

171

ransoms given—in a word, he tells us of exploits achieved by our little navy, which in themselves constitute all the "pride, pomp and circumstance of glorious war." And yet he invokes your censure upon him whom he alleges acted the part of Captain General in these heroic exploits!

Mr. President and gentlemen of the Senate:—However much I may covet these high honors for my friend, I must deal with you on this occasion with the same spirit of candor which characterizes my previous remarks. The Secretary did not command the squadron during this glorious cruise. It is true, he did apply to the President for the command, and it is equally true that he was refused; and being refused, he went only in the capacity of a private individual. To corroborate all this, I refer you to the letter of Com. Thompson, who commanded this squadron; to the evidence of Capt. Boylan, who was the second in command; to the statement of Lieut. Simmons of the *Invincible*; to the testimony of Dr. Cheesman, Surgeon of the *Brutus*. Each and all of these officers agree in the statement that the Secretary assumed no command whatever. Each and all of these officers agree that whilst on board of the vessels he was a passenger: and when he joined them in their expedition on shore, a volunteer.

Yet in the face of all this evidence, coming as it does from the highest source known to the law, the President asserts to the contrary. Upon what moral convictions his assertions are laid—upon what evidence they are founded, alas, it is not for your nor me to know. His counsel tells us with a modesty and gravity peculiar to those that represent this august personage, that upon this point his Excellency has thought proper to furnish no evidence whatsoever. "O most lame and impotent conclusion!"

But after all, let us suppose for the sake of argument, for I cannot believe you will entertain the supposition on any other terms, that the President is right, and all the persons that have testified on this point have perjured themselves; and that the Secretary of the Navy actually commanded the squadron, and is responsible for its conduct on the cruise. Where is the great sin that he has committed? Does it consist in harassing the enemy, destroying their property, attacking their towns, capturing their vessels, and erecting our standard on distant isles? If it consists in this, would to God they had sinned ten thousand times as often, and ten thousand times as much.

I know not what ideas that *august personage* may entertain upon this subject; but for myself, I fondly, proudly anticipate the day when our single starred banner shall have passed the Rio Bravo, and float triumphant through the rich valleys of Mexico. Yes, float triumphant from the glittering domes of the city of the Montezumas, and the sons of Inkleman [Inca-men?] shall have bowed before the enterprise and superior worth of the Anglo-Saxon race. For the war in which we are now engaged is a war of reformation, and the object is to promote the influence and extend the sphere of our language and our race.

The second specification in this charge consists in the capture of the British vessel *Eliza Russel*. Upon this point the President seems to dwell with peculiar pleasure and power; and this alone he thinks sufficient to work out the condemnation of the Secretary. But after a full examination of this subject, I have been unable to detect the error which the captors have committed. Whatever may be the opinion of his Excellency upon the question of international law, I know that Kent and Wheaton, Vattel, Grotius and Puffendorf lay down the doctrine that "belligerents have the right to capture the property of their enemies on the ocean wherever it can be found:" that "free ships does not make free goods." I know that publicists entertain more doubts whether neutral vessels carrying enemies' goods to enemies' ports are more liable to capture and condemnation than the goods themselves. But be this as it may, it matters not whether these great lawgivers or our President is right. The *Eliza Russel* being captured and carried into one of our ports, she should have abided the regular adjudication of the proper prize tribunal. This is law, incontrovertible law. It matters not however great the errors and outrages committed by the captors in the seizure of the vessel; they do not equal the outrage committed by the President in her release. He had no more right to do this act than myself, or any other private citizen.

I boldly affirm that the President in releasing the *Eliza Russel* transcended all bounds, usurped judicial powers and violated the laws and constitution of this land. He is the last person in the world to point out the *Eliza Russel* and cry out that an outrage has been committed. Whatever may have been the faults of others, his at least are open and palpable. It is not in this instance alone that he has erred, during his short administration. He has committed more outrages on the liberties of the citizens; more violations of the laws; more infractions of the constitution than were committed by all the Presidents of the United States, from Washington to Van Buren. And, gentlemen, whilst sitting in judgment on the conduct of the Secretary of the Navy, I trust that you will not suffer the conduct of our highest functionary to pass unnoticed.

I do not agree with the President that in the capture of the *Eliza Russel*, we have violated the rights of neutrals, nor do I fear any action which the British nation may predicate upon the conduct of our Navy in this matter: we have only availed ourselves of our established right. To conclude, I say that the President has not only usurped judicial powers, and violated the constitution of the land, but he has rendered himself personally liable to the captors and to the government for the full amount of the value of the enemies' property which he has in this case illegally released.

The President next alludes to the attempt to retain the *Eliza Russel* after his orders for her release, and the menace made by the Prize Agent, whom he alleges was appointed by the Secretary of the Navy. In reply, I will say the Secretary of the Navy was absent, and had nothing to do with it whatsoever, and that he never appointed any Prize Agent.

The President next says, that in every instance when prizes were made, that the bulk was broken, and that the most valuable articles were carried on board the vessel in which the Secretary sailed; that the said vessel was overloaded, and consequently lost, in attempting to cross Galveston Bar. I should be glad to know from what source the President derived his assurances. He has furnished us no proof whatsoever, and we must believe that his assertion is both gratuitous and false.

In his next specification, the President attempts to convey the idea that the Secretary of the Navy was callous to the sufferings of our countrymen, then in the dungeons of Matamoros, whilst he himself was doing everything to effect their release. So far as my observations extended, the very reverse was the actual fact. Perhaps, on this subject, no one felt more than I did: every exertion on my part was made for their release. I found the Secretary of the Navy a willing coadjutor. Congress with but *three* dissenting voices, passed a joint resolution authorizing the President to send our national vessels with a flag of truce, together with the Mexican officers that were made prisoners at the battle of San Jacinto, for the purpose of effecting an exchange. But he vetoed the resolution; and the Mexican officers whom we had as prisoners, were permitted to depart without conditions.

The favorable moment for making an effort for the release of our suffering countrymen was permitted to pass unseized, and they were left to their fate, to groan and languish in distant and loathsome dungeons. Had not the

President vetoed the resolution, and had I been permitted to go down in our national vessels, carrying with me the Mexican officers, I venture the assertion that I could, in the short space of thirty days, have gone to Matamoros, negotiated an exchange, procured the release of our countrymen, and restored them to their country and their friends. I say this with more confidence, because I know that negotiations were opened by Filisola with my brother for the purpose of effecting an exchange of prisoners, and that these negotiations were suddenly and abruptly broken off when he heard that the Mexican officers whom we had as prisoners had been released.

But, gentlemen, I do not wish to do the President any injustice; on this point my feelings are unutterable, and I almost doubt the correctness of my own conclusions. I therefore refer you to his veto message for his reasons for not permitting the armed vessels to go down with a flag of truce. He says that it was the danger the vessels would encounter, that caused his unwillingness to let them go. Now you will bear in mind that the vessels were then under sailing orders for a cruise on the enemies' coast; and let me ask of you, in what respect in bearing a flag of truce would their danger have been heightened? They would not be bound to come to anchor under the guns of the enemies' fort, but could have rode at anchor with safety three miles off the Bar of the Rio Bravo, or the Brazos Santiago, and opened a correspondence with the shore. Yes, I say they could have performed this with as much safety as to have cruised along the Mexican main. Whatever might have been the danger the bearer of the flag of truce might have had to encounter, there was no necessity to place our national vessels in the power of the enemy.

Perhaps in this respect, the error of the President is of the head, and not of the heart. There is at least one error which he committed that can be traced to no other source but the heart. I allude to the release of the Mexican officers. For more than a month after the capture of the *Independence*, and for some days after the arrival of Dr. Booker from Matamoros, who brought positive and certain information that her officers and crew, together with our Minister to the United States, were confined in the dungeons of that place, we had in our possession as prisoners of war, nearly every Mexican officer that was taken at the battle of San Jacinto; and they did leave this country without conditions, at that time, and under those circumstances.

I know that the President alleges that he sent an order for their detention: but if ever he did send such an order, he sent it by a careless hand, and it never reached its post of destination. The President should have sent duplicate and triplicate copies of the order; he should have sent it by so sure a channel as would have insured its delivery. I applied to one of his friends for a copy of the order, intending to send it at my own expense, but I was told that the President would attend to his own business.

Whether such an order was sent, I cannot say. One thing is certain, that in this respect the President is guilty of the most criminal neglect. It was *in his power* to have detained the officers, and *he should have done it*. But notwithstanding the release of the Mexican officers; notwithstanding I was denied the national vessels, I was determined not to be impeded by any obstacles that the Executive might throw in my way. I again applied to Congress, and that Honorable body passed a resolution authorizing the President to send a flag of truce in any way that he might please; and he finally did send me: but it was in such a manner, and under such circumstances as placed me at once in the power and at the mercy of those *inhospitable men*. When I call to mind the dangers and the perils that I encountered in that forlorn mission and when I reflect that it was wantonly imposed on me by the refusal to send the national vessels—it almost drives reason from her throne. But on this point I will detain you no longer. These things are passed—let them be forgotten. I thank God that my brother is restored to his liberty; and I thank him still more that he owes not that liberty to any effort on the part of the President.

The President next refers to a letter of the Secretary of the Navy published in the *Telegraph*, in which he speaks of a previous intention to resign. In this letter, the President says he does not manifest a disposition to consult the good of the country. But when you refer to the evidence of Governor Smith, in regard to the Secretary's intention to resign, where he says "he only remained in office for the purpose of sustaining the President, he being then threatened with an impeachment," I believe you will agree with me, that the Secretary did consult the good of the country, even at the sacrifice of his own feelings and interest.

As to the other letters of the Secretary to which the President alludes, after a cursory perusal, I was unable to detect anything criminal or even objectionable. They seemed to me to have been dictated with great spirit and genius, and to have flowed from a patriotic heart. In regard to the private correspondence between the President and Secretary which has been read to you, I have nothing to say. They are the angry productions of angry men, and I will not insult the dignity of this august body by their perusal or one solitary comment.

The President, in conclusion, says "How far the conduct of the Secretary of the Navy is right, you are respectfully called on to advise." Now, gentlemen, what that advice may be, I pretend not to know; it is a matter for which you are answerable to your country and to your God. But this much I will say:—if that advice is predicated upon the evidence before you, it will give ample satisfaction not only to the accused, but to every well-wisher of this country.

Mr. President and gentlemen of the Senate:—When I look back upon our past history; when I behold a handful of men boldly asserting their rights, and fearlessly trusting all to the God of battles, and embarking in war with a nation so powerful in numbers, and so vast in resources as Mexico; when I see our gallant MILAM shedding his life's blood upon the altar of Liberty, and BOWIE, BONHAM and TRAVIS nobly consecrating their lives to our cause; when I behold the ever lamented FANNIN and his companions dying, that Texas may be free; when I call to mind the scenes upon the plains of San Jacinto, my soul is animated, and for my country I indulge the fondest and loftiest hopes: but when I turn from this bright scene, and behold that bloated mass of inebriety and insanity, of hypocrisy, vanity and villainy; when I see him sitting like an incubus, and weighing down the hopes and paralyzing the energies of our infant republic; when I see him "Blow from his mildewed lips, on virtue blow, To blight the goodness that he ne'er can know," my soul sickens, and I turn with horror from the scene.

Here Mr. Wilson rose, and called the gentleman to order, and said he did not sit there for the purpose of hearing the President abused. Mr. Dunn also rose, and said he did not wish to hear Billingsgate language. Mr. Wharton resumed—

Mr. President:—As to the remark that fell from the gentlemen last up, I can only say that the station which he occupies precludes the idea of my replying at this time; and its total inapplication of any language that I have used, waives the necessity hereafter. As to the gentleman that feels so sensitive on the score of the President, to him I say, if I do not appreciate his feelings, I at least respect them. Would to God that the conduct of the President was such that no one could abuse him: would to God it was such that Hosannahs would be sung to his praise throughout the land. But, alas, for human nature, this is not the case. I must speak of him as he is; and after all, it is more to be regretted that he should deserve these epithets than for me to apply them. Besides, it is my imperious duty to do so; for some of these charges, charges of an infamous nature, depend on his word alone: and it therefore becomes my duty to show its worth and value.

Had he never preferred these charges, there would have been no occasion for my speaking; had he accompanied them with evidence there would have been no necessity for my reflecting on his character: but as matters now stand, I have no alternative. Gentlemen, this is a question of character altogether: 'tis for reputation we are contending. The President must not think that he can hurl his thunderbolts at other mens' reputation, and escape unhurt himself. I shall now proceed in the same manner as if this interruption had not occurred. I may, however, in the sequel of my remarks again allude to it.

Mr. President:—I said that when I looked back upon some of the bright and glorious scenes through which we have passed in our revolutionary struggle, that my soul was filled with the proudest emotions, and for my country I anticipated the proudest and loftiest destiny: but when I turn my eyes to the President, and find him an incubus weighing down the hopes of our infant republic; when I see him endeavoring to blast all that genius, and all that worth holds dear, my soul was almost overwhelmed with despair. I now warn you, gentlemen, warn you not to indulge him in his mad and wicked career. If he be not checked, he will drive from the councils of the nation every honest and worthy man, and substitute in their stead his own myrmidons: in a word, that despotism will soon be established, and the liberties of the people destroyed.

In the case of the *Eliza Russel*, he usurped judicial powers, and trampled under foot the constitution of the land.

It was but the other day that he struck from the rolls of the Army two brave and worthy officers, and confiscated their pay and bounty lands: today you are called upon to sacrifice the reputation of an innocent man, and to remove from office a high functionary: tomorrow what new victim this modern Moloch may demand, who can say? No innocence, however pure; no worth, however exalted, escapes his unhallowed touch. But I will never despair: "Virtue is God's empire, and from his throne of thrones he will defend it."

The many encroachments of the President upon the liberties of the citizen: his numerous infractions of the constitution; his stupendous usurpation of power, has awakened a spirit of apprehension throughout the land. But besides all this, there is the genius of our institutions, in the constitution of our country a redeeming hope; and I know that I will yet see the shafts of malice and calumny fall blunted and harmless at the feet of him whom they were intended to destroy. It is upon you that my hopes are founded. It is to you that your country calls to save her from the dark and deep, fathomless and hopeless abyss of despair. I cannot, nor will not believe that you, the descendants of JEFFERSON and WASHINGTON, of that illustrious band of heroes who gave light and liberty to the world, will ever sacrifice the reputation of a citizen, however humble his condition, at the command of any man, however exalted his station. Were you to do this, the shades of your ancestors would frown upon you with shame and sorrow, and you would throw your country into convulsions, compared with which all other evils and misfortunes which have befallen us would be "tender mercies."

Mr. President, and gentlemen of the Senate:—It affords me no pleasure to reflect on the conduct of our Chief Magistrate. My obligations to my friend make it an imperative duty; it is a painful task. I know that I shall incur the displeasure of his minions; I know that I shall call down upon my head the vindictive wrath of that august personage; yet I care not. "His censure and his praise alike I scorn, And hate the laurels by his followers worn."

These considerations will not frighten me from my duty. Gentlemen, this is not the first time I have known that august personage wantonly, wilfully and maliciously attempt to destroy the reputation of an innocent man. During the war, he issued an order to Capt. Baker to burn the town of San Felipe; and after it was done, finding that the measure was unpopular, he denied ever having issued such an order: leaving that brave, worthy and efficient officer not only to bear all the odium of that transaction, but perhaps to suffer the punishment due an incendiary. His unjust treatment of Col. Sherman—his persecution of Col. Coleman, to say nothing of his conduct to the lamented FANNIN, are similar instances of injustice and persecution, and will give you some idea of the extent to which that *august personage* sometimes carries his vindictive wrath.

But, gentlemen, it is not necessary for me to travel out of the record for the purpose of citing instances of this man's baseness and perfidy. During the progress of these proceedings, there was a call for the correspondence between him and the late Commodore Thompson, to which he replied in these words: "His Excellency presents his compliments to the Chairman of the Committee, and informs him that there exists no such correspondence." It was finally extorted from him; but not until he ascertained that we were about to reach it from another source.

In his message, he informs us that he has "constantly endeavored to promote harmony and concord in the departments of government." Here we find him making war upon the Secretary of the Navy: according to his own admission, he threatened to have the late Secretary of War "slicked and railed out of town." With what success this unfortunate man's efforts have been crowned, it is for you to say. You will bear in mind, gentlemen, that upon this point, the Honorable Mr. Augustin, member of your Honorable body, was recalled, re-examined and reaffirmed what he had previously stated;—merely, that it was the Secretary of the Navy, and not the Secretary of War to whom the President alluded, when he said, "he had the boys ready to have him slicked, and railed out of town."

Mr. Kaufman, who acted as counsel for the President, stated by way of apology for having recalled Mr. Augustin, that the President informed him it was William S. Fisher, and not S. Rhoads Fisher, that he threatened to have "slicked and railed out of town!" After this admission, why can gentlemen object to my reflecting on the conduct of the President? This, and this alone, attaches more infamy to the President than anything I have said, or can say. He admits that he has used such unbecoming, such scurrilous language in application to a member of his own cabinet: his admission is unparalleled in the records of falsehood or annals of man. It is a voluntary, gratuitous falsehood.

> "Here on the rack of falsehood let him lie;
> Fit garbage for the hell-hound's infamy."

I almost imagine I hear you cry out "stop, stop, stop!"

> "For nothing can'st thou to damnation add
> Greater than that."

But my duty is not yet performed. Who is WIlliam S. Fisher? And what has he done to cause the President to fulminate against him such direful threats? "He is a man, take him for all in all," I have seldom seen his like. I have served with him in the field, and in the councils of nations; he never occupied any station in which he did not discharge its duties with equal credit to himself, and benefit to his country: yet you hear him thus abused, and thus threatened.—He is absent and there lies the secret.

Gentlemen, I have endeavored to view this matter with "the calm lights of mild philosophy." I have endeavored to reason on a question of shame and honor; though from the beginning I felt the decision in my pulse: and "if it threw no light upon the brain, it at least kindled the heart." If I have not recounted to you one half of the horrid atrocities of this man that has been so appropriately termed a "demented monster," I hope that I have at least said enough to save an innocent man from the arms of this modern Moloch.

Mr. President, and gentlemen of the Senate:—To you this day is entrusted the high honor of transmitting, unimpaired, to posterity, the bright rewards of our victories, the glorious fruits of our revolutionary struggles, our RIGHTS, our LIBERTIES and our FREE INSTITUTIONS.

Thumbnail Sketch of the Second Texian Navy

With the ships of the first navy all doomed by the fall of 1837, the commodore of the fleet dead in the same year and Mexico's continued refusal to acknowledge Texas as an independent nation, something had to be done. Samuel M. Williams had been appointed by President Houston to scout and acquire ships for the Republic. When pro-navy President Lamar's term began in December of 1838, things began to move in earnest.

March of 1839 saw the arrival of the steamer *Zavala* at Galveston. The schooners *San Jacinto*, *San Antonio* and *San Bernard* arrived in the summer of the same year and the brig *Archer* arrived in the fall. The brig *Wharton* and the ship *Austin* rounded out the lineup in early 1840. The fleet found their commodore in the person of Edwin Ward Moore, a Virginia native and U. S. Navy lieutenant. Not yet 30 years old, Moore resigned his post with the U. S. Navy to accept the commission of post-captain in the Texian Navy.

The *Austin*, *Zavala* and the three schooners left for the coast of Mexico in June of 1840. The *San Jacinto* would be wrecked on this cruise. This would also represent the *Zavala's* only cruise before being laid up for repairs and then scuttled, but she was a valuable asset on this expedition. She towed the *Austin* and *San Bernard* upriver into Tabasco where Commodore Moore exacted a $25,000 contribution in exchange for not leveling the capital. The fleet returned in April of 1841.

Talks had sprung up with the Yucatan, then in a state of revolt against Mexico, to unite naval forces. For the sum of $8,000 a month, Texas would provide naval support to Yucatan. The Texian government under President Lamar agreed to the arrangement in September of 1841. The fleet

Pencil and watercolor illustration of the Texian sloop-of-war *Austin*, flying the broad pennant of Commodore Edwin Ward Moore. By Midshipman Edward Johns, 1841. In the Dienst Collection, Briscoe Center for American History.

spent time surveying the Texas coast in the summer and fall of 1841, while peace negotiations were ongoing with Mexico. With orders from Lamar to head to Yucatan, Moore sailed with the *Austin* and two schooners on December 13, 1841...the same day that President Houston was inaugurated to begin his second term. Less than 48 hours later was a dispatch sent to Moore to return immediately to Galveston. Moore did not receive the communication until March of 1842. In late March, the government of Yucatan concluded that an invasion by Santa Anna was likely a year away and that they could not justify paying for Texian assistance for so long in a time of presumed peace. Moore's fleet returned to Galveston, arriving in early May.

The *San Antonio* had been sent into port in late January with dispatches from Commodore Moore. Before returning to Yucatan, the schooner touched at New Orleans to procure provisions. On February 11, 1842, aboard the *San Antonio* occurred the only mutiny in either of the Texian navies. The mutineers were arrested and incarcerated at New Orleans.

In March of 1842, following the February attack of San Antonio by the Mexicans under Vasquez, Houston ordered the ports of Mexico blockaded. Moore left on May 8 on board the *Austin*, in company with the *San Bernard* and *San Antonio*, for New Orleans to get outfitted for the voyage. No funds were provided to accompany the orders, so Moore managed to produce about $35,000 by his own signature. Just as he was preparing to sail, the brig *Wharton* arrived in port, with no provisions, no ammunition and only nine seamen on board, with instructions to prepare her for the cruise. This would cost another $6,000 that he could not obtain with his strained credit. He immediately went to Houston to meet with the President. An appropriation of $20,000 had been made for the navy by the last session of Congress and Moore intended to return to New

Orleans with at least half of it. He advised the Secretary of the Navy that the officers of that body had not been paid for the better part of two years, that many seamen were being discharged without pay when their terms expired, that no officer in the navy had ever even received an official commission and that the *Zavala* required attention as soon as possible or she would be lost for good. Congress again met and appropriated another sum (nearly $100,000) to the navy, which President Houston signed.

While Moore was further attempting to get outfitted, the blockade was revoked and he was ordered to still go to sea and provided a new set of sealed orders. He sent the *San Antonio* down to Yucatan in hopes of renewing their former agreement and raising some cash. The *San Bernard*, laid up in Galveston awaiting repairs, was driven ashore by a storm. By November, Moore had learned that the *San Antonio* had never reached Yucatan. On October 29, 1842, Moore received a letter from the government that said, in part, "If you cannot with the means at your command, prepare the squadron for sea, you will immediately with the vessels under your command sail for the port of Galveston." This letter would become the basis for many of the charges brought by President Houston at a later date.

In February of 1843, Colonel Peraza of Yucatan arrived at New Orleans with funds and a new agreement. On February 24, Commodore Moore dropped a line to the Governor that he aimed to sail within a week. The very next day, however, the commissioners arrived, appointed by President Houston to carry out a secret act of Congress to effect the sale of the vessels of the Texian Navy. Moore won over the commissioners and one of them, Col. James Morgan, sailed with him to Yucatan in mid-April. April 30 and May 16 marked the *Austin*, *Wharton* and a handful of Yucatan gunboats engaging the steamers *Montezuma* and ironclad *Guadaloupe*. The result was an end to the Mexican blockade of Campeche and a landmark victory for wind power over steam power in battle. The Mexicans had lost far more men than did the outgunned and outmanned Texians.

The renowned Proclamation of President Houston declaring Moore and his men as pirates to be seized and brought to Galveston, meanwhile, had been published on May 6, 1843. Colonel Morgan and Commodore Moore, upon learning of this development, sailed for Galveston. When Moore arrived, he was hailed as a hero by the people, but he nonetheless attempted to turn himself in to the sheriff to answer the charges that had been made against him. The sheriff said he knew nothing of any charges within his jurisdiction and could not take him under arrest. On July 25, Moore was dishonorably discharged from the Texas Navy...without a proper court martial. Such began years of strife for Moore and the end of the Second Texian Navy.

An attempt was made in November, 1843, to sell the vessels at auction in Galveston, but citizens swarmed the auction in order to prevent any bids from being made. The vessels were placed then in ordinary, where they would remain until after annexation.

After a long 90 day trial, Commodore Moore was acquitted of the serious charges laid against him. President Houston roundly rejected the findings of the court martial and declared Moore guilty anyway. Congress disagreed and further disagreed with Houston's authority to remove Moore from command without a court martial to begin with. Edwin Ward Moore spent the next decade fighting for commissions for himself and his officers in the U. S. Navy. His feud with Sam Houston continued, as Houston was a U. S. Senator and the memorials of the Texian Navy officers were being plead before Congress. The Proclamation of President Houston always clung to him, regardless of his acquittal. A personal distaste for the navy and the

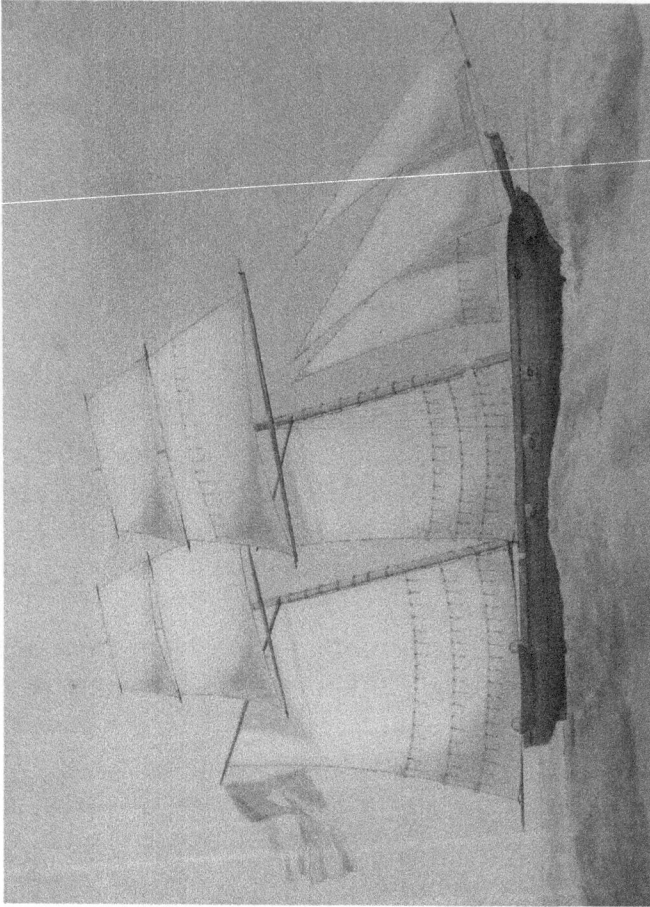

Pencil and watercolor illustration of the Texian schooner-of-war *San Antonio*, aboard which the only mutiny of the Texian navy would take place in 1842. By Midshipman Edward Johns, 1841. In the Dienst Collection, Briscoe Center for American History.

uneven temper of the Executive shut down Moore's career when it otherwise should have been reaching its prime.

The remaining vessels of the Navy (the *Austin, Wharton, Archer* and *San Bernard*) were placed in the charge of Lieutenants Tennison and Brashear until such times as the United States could take charge of them. They were sold for a pittance, owing to their varying states of disrepair. The claims of the surviving officers were settled with the U. S. government in 1857 and Moore died less than a decade later.

Midshipman George Fuller
Aboard the *Austin* in 1842

originally published as "A Sketch of the Texas Navy," TSHA *Quarterly*, January, 1904

I received my appointment as midshipman in the Texas navy with orders to report for duty on the flagship *Austin* in May, 1842. The *Austin* was then lying in the Mississippi River at New Orleans. The navy at that time consisted of the following vessels: the *Austin*, sloop-of-war, twenty guns; the two 18-gun brigs *Wharton* and *Galveston* [Archer]; and the three topsail schooners *San Antonio, St. Bernard* and *San Jacinto*. The armament of the schooners consisted of six 6-pounders and a long gun on a traverse circle amidships. These vessels were constructed by a firm of Baltimore shipbuilders and were, in beauty, speed and other seagoing qualities, unequalled. There was also an old side-wheeler, the *Zavala*, but at that time she was not in commission. Soon after joining the *Austin*, the brig *Wharton* arrived and anchored below us. The United States sloop-of-war *Ontario* was anchored a short distance further up the river, so that there was the unusual sight of three men-of-war anchored off New Orleans, as well as the revenue cutter *Hamilton*.

The officers of the *Austin* were Commodore E. W. Moore, in command; James Moore, commodore's secretary; Alfred Gray, first lieutenant; Cyrus Cummings, second; C. B. Snow, third; Wilbur, fourth; William H. Glenn, master; Norman Hurd, purser; Alfred Walker, Robert Clements, Fairfax Gray, Andrew J. Bryant, George F. Fuller, Robert Bradford, and Edward Mason, midshipmen. The officers of the *Wharton* were Captain J. T. K. Lothrop, First Lieutenant Lansing, Second Lieutenant Lewis, and Third Lieutenant Wilbur; midshipmen, Culp, White, Faysoux and Middleton. The two other ward room officers of the

Austin were Surgeon Anderson and his assistant, Surgeon Peacock.

The *Austin* was a ship of five hundred tons. Her battery of medium 24's was on the spar deck. Below were the berth deck, the steerage and the ward room. Under the latter was the magazine, under the steerage was the spirit room and purser's stores, and under the berth deck the provisions and water tanks. The bread was stowed in a locker on the starboard side of the steerage, which had a storage capacity of 20,000 lbs. of sea biscuit. The bulwarks of this ship, from the deck to the top of the hammock rail, were eight feet high, the top of the hammock rail coming flush with the top of the poop chain, and forward with the deck of the topgallant forecastle. The sleeping arrangements of the commodore's cabin consisted of two swinging cots. The ward room furnished eight staterooms for the lieutenants, surgeon and purser. All the other occupants of the ship slept in hammocks, which were swung at night and taken down to be stowed in the hammock sails in the morning.

The warrant officers in all navies are the boatswain, the gunner, the sail-maker and the carpenter, who are not officers in line of promotion. The petty officers are numerous. They are the quarter-masters, quarter-gunners, captains of the tops, captains of the forecastle, master-at-arms, armorer, purser's steward, boatswain's yeoman, and the cook, who outranks all the others.

The rules and regulations of the service were precisely the same as those in the United States navy, and copied from them, as the latter were from the English. In fact, the incidents described in the nautical tales by Captain Marryat seventy or eighty years ago might have happened as naturally on the *Austin* as on an English sloop-of-war. The daily routine was as follows: a few minutes before eight bells in the midwatch (or in land phraseology—a few minutes before 4 a.m.) the drum and fife rouse up the sleepers with reveille, immediately after which eight bells having been struck, the pipes of the boatswain and his mates are heard, followed by the cry of "All hands" and then the call "Up all hammocks." The midshipman of the watch reports eight bells at the ward room and then drops down into the steerage, shakes his sleeping successor and bawls into his ear "Eight bells! I'll thank you to relieve me."

The sailors straggle up from the berth deck, each one shouldering a hammock, which is rolled and lashed. These hammocks are handed up to the hammock rail, where they are stowed "rip rap" by one of the quarter-masters. Now pails of water, buckets of white sand and holy-stones appear, and the holy-stoning of the decks is commenced. On this thoroughly scrubbed deck, water is thrown and squilgeed out of the scuppers. The squilgee is a nautical hoe with two blades of strong sole leather. It is pushed. After this process, the deck is laboriously swabbed. When dry, the running rigging is carefully "flemished" down, that is, laid in flat coils on the deck like doormats.

The quarter-master of the watch touches his hat to the midshipman and says, "Eight bells, sir." The latter reports this to the officer of the deck, who says, "Report it to the commodore, sir." The commodore, appraised of the fact says, "Make it so, sir, and pipe to breakfast." The officer of the deck makes it so by ordering the quarter-master to strike the bell eight. It is done, and if the night has been rainy, the topsails and courses are dropped from the yards to dry. The jack rises from the bowsprit, the flag rises to the gaft, the night pennant flutters down, and the stop being broken by a sharp jerk of the halyards the broad pennant floats out from the masthead. All these movements are simultaneous.

Now is heard the boatswain's pipe followed by the single word, "Grog." The purser's steward appears at the larboard gangway with the grog-tub on which are three little tin cups. Each holds the "tod" allowed to the sailor. The men come down the larboard side of the main deck, each takes his "tod", the steward constantly refilling the little tin cups. One or two sailors, after draining the cup, turn it upside down, and if a drop falls into the palm of their hands, they rub it into their hair. To oversee this rite is one of the duties of the master's mate of the berth deck. Breakfast follows this.

At 10 a.m. the first lieutenant appears on deck. The surgeon meets him holding in his hand a small square of white paper, which has been carefully passed all around the interior of the cook's coppers. If it should show the slightest soil, the cook would receive a dozen lashes at the gangway. He was too careful, and the paper always came out unspotted. Now it is announced that the berth deck is ready for inspection. The first lieutenant descends the hatchway followed by the master's mate. He minutely inspects every crack and crevice. If he is in ill humor, he will declare the berth deck to be in a frightful condition; if amiable, he will grunt approval. There was nothing on earth so clean as the old sailing man-o'-war.

Put seven or eight midshipmen in one room, and disputes are inevitable. This sometimes ends in a fisticuff. The officer of the deck jumps down and, getting between the combatants, receives some of the blows, which the onlooker would say were intentionally bestowed. No deadly quarrels arose on the *Austin*, but on the *Wharton* two duels were engendered. Mr. White challenged Mr. Culp. They went ashore, fought, and the latter was killed.

Faysoux and Middleton met on the field of honor and, as they were both dead shots, it was predicated that they would both fall at the first fire. The result almost tempted one to believe with Buffon that chance is "the ruler of the universe." It was the fraction of a second only that saved both lives. Faysoux fired first as Middleton was coolly and carefully lowering his pistol to the mark. The bullet from Faysoux's pistol struck the hammer of his adversary's weapon and glancing from that shattered Middleton's cheekbone. As the bullet struck the lock, it caused a premature discharge of Middleton's pistol, and the bullet thus hastened by the fifth of a second, struck the rim of Faysoux's cap and came out of the top of it. Faysoux told me that he felt the bullet pass over the hair of his head. Afterwards, in Walker's time, Faysoux commanded the whole Nicaraguan navy, consisting of one schooner, with which he blew up the whole Costa Rican navy, which was represented by one brig. He was afterwards the mate of the Creole in the famous Cuban expedition, his commanding officer being no other than Lewis, formerly third lieutenant of the brig *Wharton* of the Texas navy.

Chance played another part with two officers of the Lone Star navy. The schooner *San Antonio* arrived at New Orleans to revictual and recruit. After a stay of a few days her captain made sail on a cruise after waiting a few hours for his second lieutenant who was on shore. The schooner was about half a mile away when the absent officer came down in a shore boat. The *San Antonio* was beating down against the wind. The commodore ordered the second cutter to be called away and sent the delinquent to his own vessel. The boat returned with Tennison, the third lieutenant of the *Wharton*, who had been sent on board the schooner to replace the belated man. Tennison was saved, and the other officer was lost. The *San Antonio* sailed from her port and was never heard of more. It was supposed that she was lost in the great September gale of 1843.

It is greatly the honor of Commodore Moore that he kept the navy afloat by his own credit at this time. The Republic did nothing, and Sam Houston was, and had always been, bitterly opposed to the navy. Just as Moore had succeeded in revictualizing and manning the *Austin* and *Wharton*,

197

there appeared a commissioner, Mr. Morgan, bringing peremptory orders from President Houston to Moore to sail for Galveston and abandon his intention of seeking the enemy. Morgan was talked over, and he sailed on the ship on this, which proved to be the last, cruise of the Texas navy. Everything being in readiness, a tug boat picked up both vessels one evening, passed up the river a short distance, and then turned. The crew of the *Ontario* manned the rigging and gave us three cheers, which were heartily returned and under this encouragement we commenced our cruise.

The routine of action is different at sea from what it is in port. At sea, the crew keeps watch and the "dog watch," from 4 to 6 and from 6 to 8; and changes the hours of watch each day to prevent the injustice that would be done if the same half of the crew constantly kept the mid (midnight watch.)

We left the Southwest Pass at night—a moonless night, as black as a crow's wing. I have only a vague recollection of my first night at sea on the *Austin*. The first lieutenant took charge of the deck and wore ship. The breeze was light and the motion of the vessel scarcely perceptible. When I turned out in the morning, the ship was under a cloud of canvas. The ocean was blue, the sky was blue, and all hands blue when it became apparent that the *Wharton* had disappeared. One of the afterguard coming aft with a bucket of water sang out "Sail ho!" as he reached the quarter deck. "Where away?" asked Cummings, who was the officer of the deck. "Broad off the weather beam, sir." Cummings, after ordering the man to stop where he was, hailed the lookout and rated him for his neglect of duty. As the sail was a mere speck on the horizon, he saw at once that the afterguard who sang out "Sail ho!" must be a seaman. Cummings questioned him, and the man confessed that he had enlisted as a landsman, as he would have an easier time pulling and hauling about deck than in laying aloft. Instinct betrayed him. He was put in his right place, stationed in the main-top, and proved to be one of the most effective and active of top men.

It was either on this day, or the following day, that the sentence of a court martial was carried out on board the ship. I think it was in March or April, 1842, that a mutiny broke out on the *San Antonio*. The crew rose, killed the officer of the deck, Lieutenant Fuller (son of the proprietor of Fuller's Hotel in Washington, D.C.), seized the boats, and made for shore. Six of them were captured. Four of these were sentenced to be hanged, and two to receive a hundred lashes each on the bare back. Preparatory to carrying out the execution of the decree, four lines were suspended from the foreyard after the foresail had been furled. There was not a man of the whole crew on board, from the boatswain down, who knew how to make a hangman's knot, which of course was affected ignorance. Gray, the first lieutenant, who was a thorough marlingspike sailor, exclaimed in a mildly sarcastic tone, "I'll show you how to make a hangman's knot!", which he did.

The four lines from the weather and lee yard arms, led through blocks to the deck, were married together and passed through leading blocks aft to and around the main mast and forward to a point under the yard. One half of the crew were to walk aft with the line, the other half to walk forward. The officers were all on deck, each with side arms. The prisoners were brought forth and the ropes were passed around their necks. The commodore gave the signal, a shot from the bow gun, and the crew started on their death march. The four culprits were raised to the yard arm, and must have been strangled in the ascent, for they neither struggled or made the slightest motion.

The bodies were taken down, the surgeon read the funeral service over them, and they were committed to the deep. This melancholy but necessary act of justice had a depressing effect on every man and boy on board. But the crime of mutiny accompanied with murder cannot possibly be condoned. It is discipline alone that ensures the safety of the officers, that enables them to control the crew which outnumbers them so greatly—twenty officers to perhaps three hundred and fifty sailors. I had always vowed that I would never witness the hanging of a human being, but fate compelled me to see four men hanged at the same moment.

The day following this dread execution saw the punishment of one of the two mutineers sentenced to receive a hundred lashes. The man was served up at the gangway, naked to the waist. The boatswain gave the first blow with the "cat," with its nine cords; a reddish tinge appeared as the cat was raised for the second stroke. The marks on the back assumed, as the punishment continued, a purple hue and then the blood flowed. The surgeon stood by with his hand on the culprit's wrist. At the end of fifty lashes, he made a sign that signified, "The man can bear no more," which caused his release. A shirt was thrown over his back and he was lead forward. He did not, at any time afterwards, receive the other fifty lashes, nor did the other mutineer receive one. Perhaps the commodore judged that the lesson to his crew was quite sufficient.

The commodore sailed on this last cruise a day or two earlier than he intended. He had received information that the two Mexican frigates were to re-coal at a little obscure port on the Mexican coast. For this point he made, but on arrival there found that the birds had flown. To our great delight, however, we found the *Wharton* off the place. Seen at a distance, we could not restrain our admiration at the picture she presented, with her graceful hull and raking masts—a thing of beauty, the model of a perfect man-o'war.

Once more the two vessels, reunited, headed for Campeche. In running down the coast and nearing our destination, sail was reduced, the commodore wishing to arrive in the morning and not at night. We were under three topsails, jib, and spenser. They were heaving the log when the commodore came on deck and asked, "What is she making?" "Eleven, six" was the reply of the midshipman. "A mistake, sir. Try again." A second trial showed the same result. Eleven, six knots an hour in a light breeze and under such short sail was certainly wonderful speed. The ship finally hove to and anchored off the coast of Yucatan.

At early daylight the following morning, just as "All hands up anchor" was called, the Mexican fleet hove in sight—six vessels headed by two steam frigates. The anchor was up like a flash, and our two vessels bore down on the enemy. The land breeze was blowing steadily and the enemy, miles away, were slightly to windward. At a distance of nearly three miles, the leading frigate opened fire. The shot fell short. The next one, however, passed completely over the ship. The fire was returned from the bridal port of the *Austin*, Cummings sighting and firing the gun. Suddenly the *Austin* grounded and slightly heeled over to leeward. The watchful enemy immediately got over to leeward.

Lothrop hailed from the *Wharton's* deck, "Shall I heave to?" Moore replied, "No, sir. Keep on to your anchorage." I think this cool contempt for the foe must have astonished them. We were in sight of Campeche when we grounded, and we were not much displeased to see the Yucatan gunboats coming out to our assistance, their sails white in the glistening sun, suggesting hurrying sea gulls. They came down to leeward of the ship and opened on the Mexican fleet with their long 18-pounders. A freshening breeze together with all sail packed on the ship forced her over the shoal. We bore away for Campeche with our escort of gunboats, and the enemy retired to their anchorage five miles from that city, off a little town called, I think, Llerma. This little skirmish occurred April 30, 1843.

In explanation of the appearance of allies from such a quarter, it must be said that Yucatan was in the throes of one of those peculiar revolutions so common in Mexican departments, and was in arms against the mother country. The city was blockaded and a force of 5,000 men had commenced a siege, erecting batteries in the outlying suburbs. Our arrival opened by blockade. We anchored three miles from the mole, for the land shoals so gradually that no vessel of any considerable size can approach nearer to the shore.

The city of Campeche was built by the Spaniards in the early days of the 16th century. It is a walled town, the walls being about forty feet high, with open scarp and no ditch. It was intended as a defense against the natives rather than against a civilized foe. Its battery consisted of 42-pounders mounted *en barbette*. The town had many years ago commenced clambering over the walls and sprouting into suburbs. The country about abounded in tropical fruit and here my eyes first regaled with the sight of the growing cocoa-nut. The town has no harbor, but faces an open roadstead. The Bay of Campeche, I think, contains more sharks of all sorts and sizes than are to be found in any other portion of the watery world. Perhaps this is accounted for by the extraordinary abundance of edible fish to be found in such objectionable company. The shark, however, is universally eaten in Campeche, and the fish market makes a great display of them, from baby sharks to large ones, which are sold by the pound.

We received a hearty welcome in Campeche, and the Mexicans, as if in a satirical mood, commenced bombarding the city, which kept up for three days and nights, with slight intermissions for refreshment in the way of sleep. At the expiration of the land breeze, both sides laid down their arms and gave themselves up to the inevitable *siesta*. When the sea breeze came in they resumed the game of war. The Campecheans nearly destroyed the church of St. Roman in the suburb of that name and knocked over a good many of the adobe houses. Walker, Clements and I passed a cheerful hour on the ramparts, working a 42-pounder. When we tired of this sport, we descended and were collared by three "grave and reverend signiors," who compelled us to sit down at a table under the shadow of the wall and regaled us with wine. Their manner indicated that the fracas going on was something that did not concern them, and that it would be impolite for them to interfere.

The Mexican bombardment did but little damage to the city. But one shot should have been "heard round the world" for its sportive eccentricity. A cannon ball was sent completely through the bell in one of the cathedrals, making a perfectly round hole without cracking or shattering the bell. Those who do not comprehend the exact meaning of the word concussion will say "impossible"; but there can be no concussion where resistance is not powerful enough to bring a missile to a full stop.

The first thing that Moore did after his arrival in Campeche was to ask the governor for the loan of two long guns, 18-pounders. They were sent on board and mounted on two of the carriages from which two of the 24s had been removed. "Now," we thought, "if our Mexican brothers want to play a game at long bowls, we can take a hand."

On the 16th of May, 1843, the Texas fleet set sail with the land breeze to meet the foe. They were equally alert. Moore's report has fully described this battle, and nothing remains for me but to add an incident or two. This first shot that struck the ship came from the schooner *Eagle*, the ball taking a semicircular bite out of a sailor's heel. At one time during this fight at long range, the commodore got a chance to square the yards, run between the two frigates, and engage them with both batteries. At the very first fire, the flag staff of the *Montezuma* was shot in two, and down went the flag into the sea. That ship paddled ahead and got round on the same line with her consort, to leeward of the Texas vessels. The wind died out and a short calm intervened before the sea breeze came in. The Mexicans were to leeward, but would be to windward with the coming sea breeze. The *Austin's* yards had been braced around to meet the coming breeze and, at the very first breath of it, she darted forward. Lothrop had not taken precaution against this, and the *Wharton* was taken aback. Her position was always on our weather quarter, but she lost so much in wearing that she fell hopelessly astern and could not regain her position. The consequence was that she never received a shot in the ensuing battle.

I do not remember how long the combat lasted. I only know that we chased the enemy about fifteen miles. The two steamers obstinately held their position to windward, forward of our beam. It was some time before they got the range, the shots for a long while passing too high. Their guns were 64-pounders. This was the first time guns of so large a calibre were used in action. One of these missiles dismounted gun No. 5, killed one man and wounded five others, ripped up several deck planks, and demolished a portion of the main topsail sheet bits. There it stopped and was retained as a trophy.

Andrew Bryant, a little midshipman, was struck by a huge splinter and had two large pieces of flesh carried away from one of his legs, on both sides of the femoral artery. A cartridge in its leather case mysteriously exploded and blew off the arm of the powder boy who was carrying it. The mutineer who had previously received no punishment, fell to the deck dead. His breast was a mass of bruises, but the surgeon said, "Those bruises did not kill him." I spied a spot of blood, which induced me to kneel down and lift his hair away from the top of his forehead. A small wooden splinter two inches long had been driven into his brain.

A curious experience is that which comes by being shot at from a long distance. One sees the flash of the gun, then hears the whistling of the ball, and then the report, the ball out-traveling the sound. After a little study of the coming balls one could determine very nearly where they were going to strike. Two of them I shall always remember. Of the first one I said, "This is going to pick a man from my gun's crew." It struck just under the port between wind and water. As it was jammed between two of the timbers, it was found impossible to drive home a shot plug. The other shot which announced its intention to become intimate struck the deck of the topgallant forecastle directly over my head (for I was at gun No. 1), and tip-tip-tipped overboard, simply denting the planks. Walker, who was master's mate of the forecastle, looked over and, with his peculiar lisp, exclaimed, "Fuller, that was devilishly close!"

A few moments later, I heard an oath from the sail maker who declared that the scoundrels had ruined his new jib. It was of light raven duck. A cannon ball had passed through it and, the wind freshening, was reducing it to ribands. Now, some of the standing rigging having been shot away, together with a good deal of running rigging, the commodore wore ship to take the strain off the starboard and to engage the enemy with the larboard batteries. This heeled the starboard over to such an extent that she made water rapidly through that shot hole under gun No. 1 and absolutely compelled a return to our anchorage.

The *Montezuma* was so crippled that it was twenty minutes before she could stir. The *Guadalupe* followed after us for a short distance and then turned back to her consort. The sailing vessels of the Mexican fleet fled at the moment the sea breeze came in. The most unaccountable mystery connected with this flight in which a superior force, more than three to one, fled from their adversaries, was the inaction of the Yucatan gunboats, which obstinately remained at anchor defying all signals made by Moore for them to make sail. At the last moment, one of them, commanded by a Frenchman, came out and opened fire on the foe. The commander of the gunboats, an American by the name of Bowie or Bowen, came on board when we anchored, looking frightened and deadly pale. What kind of rating he received from Moore, we never knew. In this action, the fire of both the *Austin* and the *Wharton* was directed exclusively at the *Montezuma*, while both steam frigates directed theirs exclusively at the *Austin*. Neither the *Guadalupe* nor the *Wharton* received an accidental shot.

Soon after this battle, we received information that Sam Houston had issued a proclamation denouncing Moore and his men as pirates, calling upon all the nations of the earth to seize us wherever found on the high seas. This was an atrocious act to be done by an otherwise honorable man and brave soldier. We had unwittingly been fighting with, figuratively speaking, halters round our neck.

Years afterward, when I was in company with two officers of the United States navy, Houston's extraordinary act was discussed. One of them said "I was attached to the *Vincennes* sloop-of-war at that time, and we had received orders to seize the Texas vessels and bring them into port for their protection."

"Supposing," said the other officers, "they had declined your generous offer?"

"We should have captured them—a very easy task, I fancy."

"Much more difficult than you imagine," said the other officer. "I knew the *Austin*, her officer and crew. She was undoubtedly the fastest sloop-of-war in the world. She could have sailed round and round the *Vincennes*, raked her, dismasted her, and left her a wreck on the water. In addition to her great sea qualities, she was well commanded, well officered and well manned. You would have scratched a Russian and found a Tartar."

When it was known that we were to sail for home, it was reported that Marin had sent word to Moore that he need not hope to leave the coast alive, and that Moore had replied that he would be happy to meet him outside, hoping that he would find courage enough to come to close quarters. I know not what truth there was in this, but I know that on the morning we got "up anchor for home" the Mexican fleet had disappeared from sight.

On the 4th of July, we anchored off Sisal. On shore two of us celebrated the day with an omelette of sea turtle eggs. These little round yellow shelless balls which the turtle lays in the sand would defy the digestive powers of an ostrich. A day or two afterwards, we anchored off the Alacranes and obtained a stock of sea turtle, at which Jack seemed inclined to turn up his nose and bawl for salt beef and pork. Our next anchorage was in Galveston harbor, which seemed full of boats filled with people who uttered shouts of welcome. We were given a public dinner by the citizens, and here ended my connection with the Texas navy, and with Texas, which country I have not seen for nearly sixty years.

It seems a pity that the Republic of Texas ever became merged in that of the United States. Today, she would have been an empire in herself, with the power to buy and sell where she pleased. Strange as it may seem, there are only a few people in New York City who know that the State of Texas was ever a republic, so very few that I have yet to count the first one. I do not know whether the old flag of the child Republic is now the State flag, but I hope it is, and that it may wave until the last syllable of recorded time.

Pencils sketches from the journal of Midshipman Edward Johns, a New Orleans native who served aboard the *San Antonio* and *Austin* in the second Texas navy, 1841-1842.

CORNELIUS COX OF THE TEXAS NAVY

originally published as "Reminiscences of C. C. Cox," TSHA *Quarterly*, October, 1902

My parents began housekeeping about the year 1812. They were both born in Kentucky. My maternal grandfather emigrated to that state from Maryland. He served in the War of the Revolution and moved to the "dark and bloody ground" in the days when it was necessary to carry his rifle in the field so as to be always prepared if a red man called. This grandfather of mine lived to the age of 94 and died beloved of all men. I had the honor of inheriting his Christian name, but his virtues have fallen by lightly upon my shoulders. My dear mother still lives.

My father's ancestors moved from the "Mother of States" with the early pioneers who crossed the Ohio River. The earliest account I have of this grandfather, he was engaged in a large saddlery business in Lexington. My father was brought up in that trade and followed it to his grave. His life was an eventful one, saddened by many misfortunes but cheered also by happy surroundings, and his great loving heart and genial disposition were perpetual sources of joy to him and pleasure to his friends. At 88, he crossed over the Dark River and memories most dear follow after the dear old man.

Of brothers and sisters, I will have occasion to speak hereafter. Suffice it to say now that there were five of us in all—3 sons and 2 daughters. The two last and the elder brother have long since gone to the spirit land. The baby brother yet lives and dispenses hospitality at the old paternal home, and the other brother, well it is of him and his career that I am now attempting to write.

In the little town of Piqua in Ohio on the 7th day of December, 1825, I am told that my eyes first opened to the light. At the age of 5 years, I was so far developed mentally and physically as to be able to accompany my parents on their return to their native state. From this time I can date my earliest recollections, and in the succeeding seven years are comprised the incidents and pleasures of my school days, stick horses, skating, and first love. In looking back now over the lapse of fifty years, much of that part of my life is as vivid as the occurrences of a much later period.

Benedict Knott taught the school at the "Forks." He was a "Tartar" among the boys. I do not remember to have seen him whip a girl, but he feasted on boys. He was succeeded in the school by Mr. Samuels, a much milder man, but pedagogics in those days were practiced from a different standpoint than prevails in modern schools. Dogmatic authority and apple tree coercion were the methods employed to develop the juvenile intellect. A comparison of that system with the methods and discipline in vogue today, with the fact in view that children have been the same in all ages, is calculated to increase our respect for the primitive plan. I had to walk 3 miles to this school. I say walk, though I generally trotted or galloped—not astride of a real horse but straddling a stick horse—and it seems to me now that the pleasure of that exercise is inferior to none that I have ever experienced.

Our home was situated upon a beautiful hill overlooking Elkhorn Creek. For a hundred years that has been the one home in our family—the Mecca to which children and grandchildren, though scattered to the winds, have periodically journeyed to recreate, to recuperate and to enjoy the glad welcome of parental love and hospitality. I have wandered much over this fair land of Uncle Sam's but I have never seen a spot in all its length and breadth that was so beautifully possessed of all nature's choicest gifts as the land of Boone.

My eldest sister married Sidney Sherman some time in 1835. His history I need not give, for it has already been written. It will be remembered that in this year began the struggle between Texas and Mexico, which culminated in the Battle of San Jacinto on the 21st of April 1836. In the fall or winter of that year, Colonel Sherman returned to Kentucky for his wife, and in the month of December following, he moved to his adopted home in the Lone Star Republic. It was my fortune, good or bad, to constitute one of his family from that time. I was just twelve years of age. My school was ended, my home abandoned and my future life and prospects in the bosom of Texas buried.

The journey to Texas was devoid of any special feature, except that to one so young and who for the first time was viewing the busy world, the incidents and scenes encountered on the trip were as novel and fascinating as the shifting objects of a panorama. From Frankfort to Louisville, we traveled by the pike, the beautiful snow falling thick and fast on the day of our departure. At Louisville, we embarked upon the splendid river packet *Henry Clay*, commanded by our friend and neighbor Capt. Jack Holton, one of the old time Kentucky gentlemen. Our company was augmented by a Mr. Humphries and his two sisters, also en route for Texas. Mr. H. afterward became a member of President Lamar's cabinet.

In due course of time, the boat arrived at New Orleans and there we transferred to another steamer. We proceeded up one of the small bayous of Louisiana as far as the stream was navigable and here we disembarked, and thence completed the journey to Houston overland. I say overland, but anyone acquainted with western Louisiana and eastern Texas in the winter season will know that we had as much water as land on that trip. They will also realize the fact that to ladies who were taking their first experience in camp life, and still tasting the comforts and good things they had left behind, the conditions which accompanied us throughout this part of the journey were not calculated to reconcile them to the sacrifice they had made.

Our party were, however, well prepared for traveling. Mr. Sherman had brought with him a comfortable carriage and several find Kentucky mares and horses. Humphries also had a suitable outfit. To our party also belonged a very handy young Mexican that the colonel had kept with him in the capacity of servant since the Battle of San Jacinto. He called himself Francisco and claimed to have been Santa Anna's bugler. He remained with us several years until an opportunity was offered to send him to his home in Yucatan.

Colonel Sherman settled on San Jacinto Bay at a point about midway between Galveston and Houston. This was near Morgan's Point and about 7 miles below the battle-ground. The first little place he occupied was called Mt. Vernon, a very pretty site for a house, being on a bluff overlooking the bay. Here we spent about a year. The house was built of logs and contained just one room—but that room was either very large or stood cramming remarkably well.

Shortly after we had located, the family was increased by the arrival of an aunt, niece and nephew of Mr. Sherman, and a little later on Mr. Dana Sherman and his wife arrived. With these accessions, we had 9 in the family besides the cook. I don't know where she slept, but certainly not in the kitchen, for that family convenience was just outside the door without other protection than a little brush overhead.

I don't know why it is that everybody wants to keep a hog. An old sow can do more mischief and cause a man's wife more unhappiness than all the other animals on the place, if you have no yard fence and the kitchen is not walled up. We had a sow, and I have seen her take the lid off the oven and appropriate the contents when the fire was hot—and manifest not the slightest remorse at the freedom of the act or the least sensibility to the warmth of her repast. This was one skeleton in our house, but as there were no closets, we had not room for many such.

We had a dog also, but he was a noble fellow, a Newfoundland, loved the water as if it were his native element and if not a regular Nimrod, was certainly his shadow. Ducks, geese and swans almost literally covered the waters. The deer came in sight of the house in droves—and fish at the bay shore in variety and abundance. Cattle were plenty and cheap, and we had Kentucky stock to ride. Only one neighbor within 2 miles. This was as near as I can remember how matters stood with us the first winter in Texas.

It may be proper now to say something of the personnel of the people who preceded us to this part of Texas and who now constitute the citizenship of the Bay Country, for it must be remembered that Texas was very new, and at this time very sparsely settled. Except for the occasional settlements, the country was in a state of nature. The savages that erstwhile had held the land had been driven towards the frontier but there remained some remnants of the more peaceful tribes to remind us of the late sovereignty of the Noble Red Man. The families settled along the bay shore, on either side, were mainly from different southern states and came in with Austin's Colony—each head of a family having received a headright of one league and labor of land, being near 4,600 acres. These families were generally living upon their own locations and consequently neighbors were usually 2 to 3 miles apart.

We had one near neighbor, say within half a mile. This was Enoch Brinson, who had emigrated from Louisiana. There was always a mystery about this man. He had lost an eye and always wore a large tuft of hair over where the eye had been and always kept his hat on his head, even at meals. At the time I write, he was about 50 years of age. Mr. Brinson was a very social, hospitable man and an obliging neighbor. He was a hardshell Baptist of the ultra kind—predestination and all. His wife was a good little woman and one of the sort that never tires. She usually milked 30 to 40 cows night and morning, and supplied the family, from butter and cheeses to chickens and eggs, that she marketed in Galveston. And here comes in another member of the family, though neither kith nor kin, but sort of a silent partner, who did all the chores and outside work and ran the boat that carried the surplus to the Island City. This was Mr. John Iams, bachelor and friend of the house. John was a good, honest fellow, and clever, handy and full of fun on all occasions. Fortune had not blessed this home with the prattle of little ones when we first knew them, but later on there came to them a boy and a girl. Mrs. Brinson had a charming little niece whose visits from across the bay were always much enjoyed by me.

The next neighbor down the bay shore was also a bachelor, and more of a character than the rest. This was General Clopper, but as we are shortly to be nearer neighbors to the general, I will reserve him for future mention. Two miles further down the bay, we come to Morgan's Point or New Washington, the home of Colonel James Morgan, a participant in the active scenes of the late unpleasantness with Mexico and at the time of the Battle of San Jacinto was in command at Galveston Island. The colonel was the agent and active partner of northern capitalists who had invested largely in Texas lands, under the name and style of the "New Washington Association"—and hence the name of the colonel's residence, which at that day was the most pretentious dwelling in all the land. The situation was not only the most prominent but the most beautiful site on the bay, being at the junction of San Jacinto and Galveston Bays it overlooked both waters for many miles and, though somewhat bleak in winter, it was a delightful location in summer. This home was highly improved and exceedingly attractive.

The family was cultivated and hospitality was spread with a lavish hand. The colonel's wealth and social and political prominence in the state, his liberality, genial disposition, love of company and fine conversational powers made his home the resort of the stranger as well as friend. I have met at this house President Burnet, General Houston, Sidney Johnston, Barnard E. Bee, Dr. Ashbel Smith, Mosely Baker, Mr. Anderson (the naturalist), the son and daughter of John Newland Maffit, Commodore Moore and hundreds of other prominent citizens of Texas and other states who came to this home to enjoy its comfort and hospitality. The old colonel has long since been gathered to his fathers.

I scarcely know how to write of the next ten years of my life. Such a multitude of incidents crowd upon the memory that I know not which to speak of and which to pass by. Anyhow, the bay was home to me in all these years, but much of the time I was far away in body, though in spirit was roaming the beautiful prairies or sailing upon the lovely waters. In only the second year of our residence on the bay, death entered into our household and carried off both Mr. Dana Sherman and his wife. They died the same day of yellow fever. A beautiful little daughter was left us as a legacy.

About this time, I was placed in a store in Houston, my brother-in-law thinking it would compensate me in the loss of other schooling—but the venture was not productive of any great amount of good. The firm was about busted when I entered the store and, in a few months, closed business. However, I had some experiences which I can never forget. Mr. Neighbors, afterwards Major Neighbors the Indian Agent, was a clerk in the store. He kindly took me to room with him or, rather, to sleep with him, for our quarters were in a loft in a building apart from the store, and our bed a few blankets on the floor. Such accommodations would have been satisfactory but for other company. The fleas were as thick as the sands of the sea. Our clothes were actually bloody, and our bodies freckled, after a night of warfare with the vermin. And the rats—I cannot convey an idea of the multitude of rats in Houston at that time. They were almost as large as prairie dogs and when night came on, the streets and houses were literally alive with these animals. Such running and squealing throughout the night, to say nothing of the fear of losing a toe or your nose, if you chanced to fall asleep, created such an apprehension that, together with the attention that had to be given to our other companions, made sleep well nigh impossible.

We boarded at a hotel near the Bayou, and I can almost smell the dining room yet. In those days, the markets did not furnish fresh vegetables. But onions in barrels and boatloads were everywhere and in everything and the smell of onions and the taste of onions followed us day and night like a nightmare.

But I do remember one pleasant dining during my short sojourn in Houston. I was loitering upon the street in the vicinity of the principal hotel, when my hand was suddenly taken by General Sam Houston—and with gentle condescension, this wonderful man strode into the hotel. Reaching the dining table, I found myself at once seated between the Ex President and the then President of the Republic of Texas, General Mirabeau B. Lamar. It seemed to be a special dining as the company was numerous and select. And to say that I was stunned and almost paralyzed by this presence would not do justice to my feelings. It seemed to me that the company regarded me as the distinguished guest. I felt that all eyes were upon me—and the shots continually fired at the General and his protégé covered me with confusion.

The wine flowed freely, and when a toast was drank, my glass had to go up with the rest. In this way I suppose I gained my self possession. At any rate, I left the company with a feeling of enjoyment, and the memory of that little compliment from so distinguished a source has always been a pleasant remembrance.

When I returned to the bay, Mr. Sherman had removed his residence to Crescent Place, a point on the bay two miles above New Washington. This was in 1839. In that year the wheel of fate made another revolution on my account. My brother and sister, ever anxious about my educational necessities, thought they saw a solution to the matter in the opportunities offered in the Texas Navy. Accordingly, an appointment was obtained for me as a midshipman, and orders furnished for me to report to the commanding officer at Galveston.

Now at this time, the city of Galveston was not the attractive place that it is fifty years later. The population probably didn't exceed 2,000. The houses were plain wooden structures ranging from the little 10 x 12 shanty to the somewhat pretentious, storehouses, here & there a respectable looking dwelling, and of course the indispensable hotels, which were ample for the needs of the town. The wharves, which in later years have formed a bulwark for the city from the storms and waves that come down from the north, had not been built. On the occasion of my first visit, the steamer ran head-on to the shore, or as near as the water would allow, and the passengers disembarked on staging from the boat to the shore.

The storm which had swept over the Island in 1837 had left many reminders of its visit. One schooner was embedded in the sand just where we landed. I saw another at the sand hills over on the Gulf side of the Island. But our new Navy rode at anchor in the harbor and made cheerful the otherwise gloomy prospect.

The brig *Wharton* and schooners *San Jacinto*, *San Bernard* and *San Antonio* were in port when I went down. My orders were to report on board the *Wharton*, which vessel was under sailing orders for New York. But when I presented my papers, the *Wharton* already had her complement of middys, and I was assigned to the schooner *San Jacinto*.

And now began an experience and mode of life for which I soon discovered I was not intended. Our lieutenant commanding was a man by the name of Gibbons, the most tyrannical officer that I have ever known either in the navy or army. Some of our men were real land lubbers and of course had to be drilled in the duties of the ship. But to run up the rigging and out on the yard arms, and swing yourself like a monkey by one hand or balance yourself on the foot rope forty feet in the air and furl and unfurl sails like an old Tar was just not what the recruit could do. But the lieutenant had great faith in the "Colt" and for every blunder, poor Jack would have to come down and lay himself across the gun and receive a dozen from the boatswain's mate.

Well it was not long after I went on board until our vessel was appointed to service. A schooner loaded with army supplies was ready to sail from Galveston to Velasco at the mouth of the Brazos, and our man-of-war was ordered to convoy the schooner down. So one bright sunshiny morning, our schooner was taken in tow by a steamer and carried outside the Galveston bar, to there await the sailing of the merchant schooner. But for some purpose not now recollected, two small boats belonging to our vessel were left behind with orders to follow on later in the day and join the ship outside. Each boat was manned with four men with a midshipman in charge. I was in charge of one of the boats. Now, as everyone may not understand the iron rules of the naval service, the relations of officers and men, and the discipline that is observed on shipboard, let it be understood that here was a boy not yet 14 years of age, who had never tasted of salt water, without judgment or experience, suddenly clothed with the dignity and authority of a commander—a mere infant in intelligence but a very Titan in authority. And now, after all these years, in penning these recollections, I am oppressed with shame and mortification at the abuse of the position I occupied, and the want of consideration and respect for the feelings and gray hairs of the old tars that composed my crew. I was but an infant upon the waters. They were veterans of the deep. But then I was a little officer and they were the machinery that propelled my boat.

Well, as I have said, the morning was beautiful. They bay was as smooth as a lake and scarcely a breath of air was to be felt. About 10 o'clock I pulled out from the city and a few minutes later the other boat followed. We were not long in reaching the east end of the Island and, on turning the point and heading for the bar, we soon encountered a heavy sea coming in from the gulf. And now a dense fog settled upon the waters. Still, we kept on. Out upon the bar, the seas rolled not mountain high but so high that our little boat danced among the waves like a toy. The men said it was madness to go on, that we would be swamped, and we had best go back and wait for the fog to clear away and the sea to abate. The other boat did go back, but I had orders to join the ship without delay, and I had not the courage to disobey any order of Lieutenant Gibbons. So we pulled ahead, head on to every wave, the spray dashing over us with every pitch of the boat and without compass or objects to guide us.

It is now about 12 o'clock. A very little breeze is springing up, and right ahead not fifty paces distant, we descry a vessel under full sail outward bound. A few lusty pulls brought us alongside the stranger and no boarding party ever reach the decks of an enemy with more alacrity than myself and my men on the deck of that vessel. The captain treated me kindly. He was bound for Mobile. He refused to lay-to until the fog cleared away. He fired off his gun and blew his horn to attract our vessel, if in hearing distance, but no answer came. In the meantime, the vessel was slowly going seaward. The captain said I could stick to him or take my boat again, but now his dinner was announced and he invited me into the cabin. I thought I was hungry and took my seat at the table with great willingness. Pork and beans occupied the center of the table, or that dish seemed to have more prominence than all the rest. My plate was helped and I got a piece of pork in my mouth. But just then I found difficulty in swallowing. The cabin seemed too close. A cold sweat began to break out on me and, excusing myself to the captain, I returned to the deck in double quick time, and there delivered my first tribute to the Old Ocean.

I was dreadful sick, but I had not long to indulge this weakness. Our friend the schooner was gliding along lazily in the fog, and about 1 p.m. we hove in sight of a vessel at anchor. This was a brig loaded for Galveston. Being now out over the bar, the sea was not so rough. Thanking our friends of the schooner for their hospitality, we reentered our little boat and pulled for the brig. About 3 o'clock the fog cleared off and enabled us to see the *San Jacinto* about two miles off in the direction of the Island. I pulled alongside about 4 o'clock and, mounting to the deck and touching my cap to the lieutenant, I briefly explained the cause of my delay and was rewarded by a reprimand for my temerity in pulling out to sea in a fog.

The following morning found us riding quietly at anchor, ready for sea, but waiting for a breeze. The sea was almost smooth and not a breath of air astir. The crew was practised at putting on and taking off the sail and other maneuvers in handling the ship. At 12 o'clock, when all hands were piped to dinner, the sails were left spread and a peaceful stillness pervaded the vessel. Just then a windy visitor came upon us with such suddenness, force and fury that before the lieutenant commanding could get on deck, the schooner was lying on her starboard side with the foresail and mainsail in the water. Lieutenant Gibbons shouted, "Let go the sheets! Let go the hallyards!" but the men seemed paralyzed and only after repeated orders and by his own efforts were the sails so lowered and shifted as to be relieved of the force of the wind. Then the schooner slowly righted and faced to the wind. The sails were rapidly taken in, the anchored weighed, and we drifted off before the storm. The norther was a terrific one. We lost the vessel we were conveying and, on the fifth day, pulled up at the mouth of the Brazos, only to discover that she was already safe inside the harbor. In a few days we were again lying in the harbor at Galveston.

I have been somewhat tedious about this, my first trip to sea, simply because it was my first. It was a very short expedition, and without incident except for the storm, but it gave me a foretaste of sailor life and, since I was seasick the greater part of the time, my first impressions of riding upon the deep blue sea were not the most agreeable.

I will undertake to follow the daily events of my brief service in the Navy but will give the prominent features in a few words. From the *San Jacinto* I was transferred to the steamship-of-war *Zavala*, Captain Lothrop commanding. This vessel carried about ten guns and was a well equipped man-of-war. When the ship left Galveston, she proceeded to New Orleans. Here we remained a short time, enlisting men and taking in supplies. Thence we proceeded on a cruise in the gulf and after some days anchored at the Arcas Island, not far off the coast of Yucatan. Among the recruits who joined us at New Orleans was a young midshipman, whose name I have forgotten, who had contracted yellow fever and was taken down soon after coming on board. I do not now if I took the fever from him, but I do recollect that I was sick and that we lay together in the saloon of the steamer. The young man died at my side.

I do not remember how much times we spent at the Islands—perhaps a month or two. But in the course of time, our vessel appeared off the mouth of the Tabasco River and came to anchor about sundown one evening, it being then too late to cross the bar. The sea was quite smooth, the sky clear, and not a breath of wind. Very soon a heavy sea came rolling in from the gulf. The strong current from the river which, after entering the gulf took a course along the land, made the ship ride in the trough of the sea, and she rolled from side to side like a great log. Orders were at once given to weigh anchor and get underway, but before that could be done, a huge wave carried away our rudder. This rendered us helpless and the order to get up anchor was countermanded.

Now commenced an experience the likes of which I expect few sailors ever witness. We lay in this position for five days, with no wind but the waves rolling in mountain high. We were about two miles offshore, our anchors dragged some and the vessel sometimes gave a heavy thump on the bottom. To lighten the ship, our guns, one after another, went overboard—the shot had gone over first. We cut away the masts so that the ship would not be so top-heavy. Our coal gave out, for we had steam up all the time, and the bulkheads and available parts of the interior of the ship were cut out to make fuel.

In all these days and nights, the vessel rolled like a log—first one wheelhouse then the other, under water. It was unsafe to be on deck without fastening yourself to something. Every moment it looked as if the next would upset the ship or knock her to pieces. I was dreadful seasick and felt quite indifferent to danger. The morning of the fifth day, the sea subsided. We got up anchor and with an improvised rudder, steamed over the bar and up the river five miles to the town of Frontera. Our handsome steamer was almost a wreck.

Yucatan at that time was at war with the central government of Mexico. Texas and Yucatan were in alliance and
our fleet was ordered there to aid in an expedition against
the centralist troops who were in possession of the city
of Tabasco, about eighty miles up the river. We were the
recipients of much attention while at Frontera. The *Zavala* was the first steam man-of-war ever seen in that river.
Hundreds of people, ladies and gentlemen, came down
from Merida, the capital of the state, to visit the ship.

Here I was taken with the scurvy and had a lingering spell
of sickness. As soon as I could be moved, I was taken on
shore and nursed by a good lady of the place. Other vessels
for the expedition appeared and the fleet steamed up the
river in tow of the *Zavala*, and all under the command of
Commodore Moore, whose flagship was the sloop-of-war
Austin. But we had no fight. The enemy vacated the town
before we reached it and after one night's stay, we again
dropped down the river, but a good many bags of silver
were taken on board our vessel at Tabasco and a good portion at least was distributed among the officers and men
of the fleet as prize money. I think eight dollars was the
share I got.

This about ended my active service in the Navy. On the
return of our vessels to the Arcas Islands, I was transferred
to the *Austin* and, after a short cruise in the gulf, she entered the harbor at Galveston. And now, after something
over two years' service in the Navy, with no prospect of
active service in the future and finding that I had neither
taste nor fitness for the life, I resigned my commission and
returned to the home on the bay.

Before taking a final adieu of this period of my youth, I
must indulge in some other reminiscences of the time and
incidents connected with my sojourn in the Navy. The life
is a hard one, the discipline rigid and a boy of 14 and 15
has not the physical capacity to perform the regular watch

on shipboard—four hours on duty and eight off—with two "dog watches", 4 to 6 and 6 to 8 p.m. each day put in to alternate the watches. In case of dereliction of duty, the usual punishment for an under-officer is double duty, that is four hours on and four off. On one occasion, tired nature dropped me into the arm of Morphius, when I should have been walking the deck. This was death by the regulations. Lieutenant Gibbons commuted the punishment to double duty for two weeks. In discharging the sentence, I forfeited my life several times, but as it was necessary for the offense to be discovered before punishment was inflicted, I escaped hanging, always by a timely warning. On each of the vessels that I served, I was favored and befriended by the lieutenant in whose watch I was placed. I must ever feel grateful for the kindness and generosity of Lieutenant Tennison of the *San Jacinto*, Lieutenant Seegers of the *Austin* and Sailing Master Baker of the *Zavala*. They treated me like they were my older brothers.

When one of the seamen committed an offense or violated an order, the punishment was frightful. Flogging with the Colt was a common pastime, a daily occurrence, a sort of misdemeanor penalty. But graver offences were rewarded with the "Cat of Nine Tails." Three dozen licks on the bare back was the usual dose. The culprit stood at the gangway with his hands lashed to the rigging, his feet fastened to a grating on which he stood. The man was stripped to the waist, all hands on deck to witness the scene. The articles of war were read, the ship's physician on one side and the boatswain on the other. When all was ready, the flaying commenced. At each stroke of the lash, the solemn count, 1-2-3 and so on, was proclaimed aloud and the poor criminal would cringe and grunt at every blow. By the time the three dozen, the usual complement, were given, the fellow's back was variegated with the colors red, black, blue and white, the blood running in little rivers at his feet. It is gratifying to know that this barbarous practice has been abolished by most of the nations of the earth.

Burials at sea are attended with the same solemn character as the interment of the dead on shore, but the procedure is different. The corpse, after being dressed, is then sewn up in canvas, with two round shot at the feet, then placed on a plank reaching out over the side of the vessel. The entire crew are piped on deck, the burial service is read and then the end of the plank is raised and the departed goes off into the sea foremost. The same rites are given the criminal who is hung at the yard arm.

Texas was poor in that day and could not furnish her pantries with many delicacies. Salt beef, salt pork, beans, tea and hard tack were the staples. Our crackers were nearly always old, musty and full of worms. The worms were easily disposed of by heating the bread and then knocking them out, or soaking the crackers in hot tea. They were easily killed and I never discovered any difference in the taste of the worms or the bread. On one occasion, our vessel was furnished with a lot of chocolate beans purchased at Campeche, which we roasted, ground and used as a substitute for coffee.

The daily life on a ship is monotonous, but the sailors have their pastimes and employments when off duty. They wash, mend and often make their clothing, especially hats. I made myself a straw hat and one pair of pants while in the service, and had my arms tattooed as all old sailors do. Our vessel, the *Zavala*, laid at the Port of Sisal a good long time. It was here, I think, instead of Frontera as before stated, that so many ladies came to visit the ship. I thought the Mexican girls beautiful. They all smoked and each carried a little bunch of cigaritas. The etiquette was to place a cigarita in the mouth, light it, and then hand it to the other party. Such a temptation very few young men can resist.

If I could do justice to the subject, I would like to tell more of the Arcas Islands. As well as I remember, there are three small islands, set in a triangular position, with a small but beautiful body of water in the center, which affords safe harbor for vessels drawing 20 to 30 feet. We anchored in about three fathoms water, probably 200 yards from shore. The water is very clear and objects on the bottom are distinctly seen. Here we had fine fishing and a species of fish abounded that I have never seen elsewhere called the parrot fish—the head and half the body was a bright green, the balance of the fish the usual color. Some were quite large, weighing ten to twelve pounds. Along the shore, sharks were numerous but we saw none out at the vessel. The islands furnished many varieties of shells and we collected beautiful specimens of coral. It was delightful bathing in this saltwater lake. Commodore Moore was the best swimmer I ever saw. He could float like a feather on the water, and swim on his back as fast as most men can the ordinary way. I have seen him leap from the top of the wheelhouse of the *Zavala*, some 20 feet above the water, and go to the bottom, a run of about 40 feet.

There were no trees and almost no vegetation on the islands, but they are a great resort for the birds of the ocean, and we captured a great many eggs and young birds to eat. But now I have done with the sea and for some years to follow, would be found growing and ruralizing at Crescent Place on San Jacinto Bay.

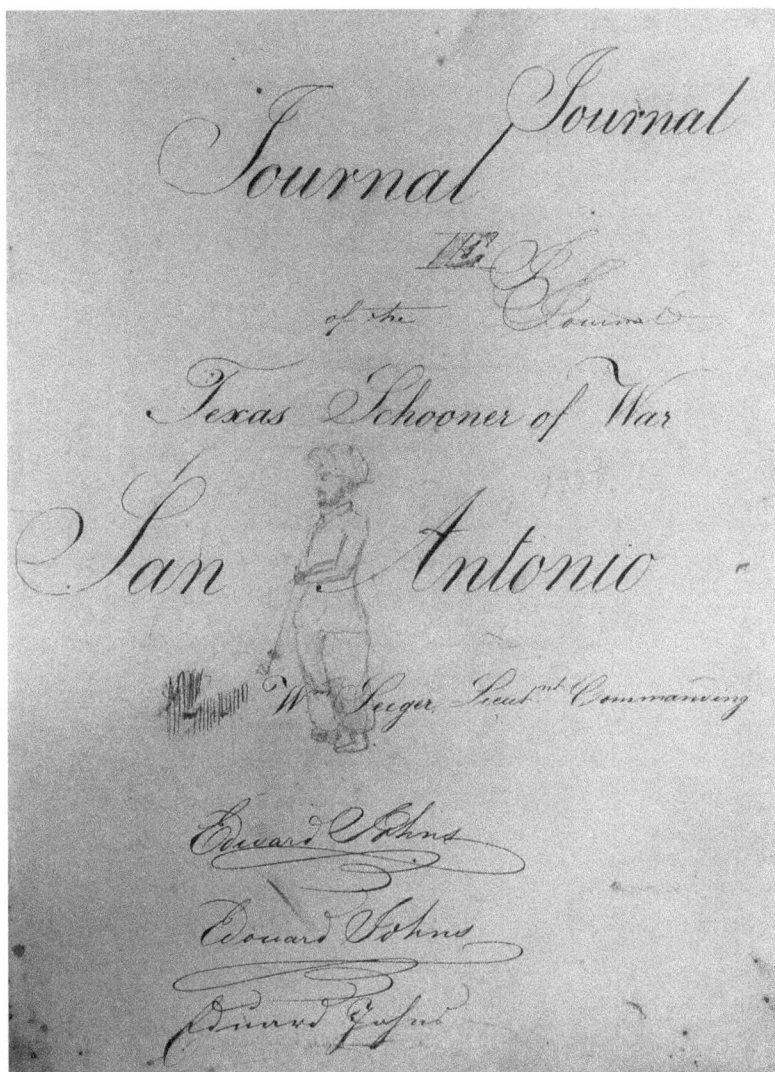

Illustrated title page of the Edward Johns journal, Dienst Collection, Briscoe Center for American History.

Texas Independence Day salute, 1842. Edward Johns journal, Dienst Collection, Briscoe Center for American History.

Journal of Alfred Walke,
April 16-27, 1843, Aboard the Austin

Journal of a Cruise on the Texas Sloop-of-War
Austin, Commodore E.W. Moore, Commanding

1843

Remarks Sunday April 16th 1843

At 9:15 p.m., we got underway in tow of steamer *Lion* down Mississippi River. The U.S. sloop of war *Ontario* gave us three hearty cheers which was returned by the crews of both vessels (this ship and the brig *Wharton*, our consort) & after several delays caused by fogs. Trying patent shot invented by Dr. Maxey of New Orleans and the tow boat leaving us at the Balize did not arrive in salt water until —

April 20th

At 5:45 p.m. we weighed our anchor in company with the brig *Wharton*, made sail, beat to quarters, shotted the guns. Beat the retreat and steered for Telchac when we expected to meet the Mexican steamer *Montezuma*. Latitude by observation 28°50' N at Meridian.

April 21st

During the night of the 20th, the brig parted company with us. Exercised our crew at general quarters from 4 to 6 p.m. At 10:30 a.m. called all hands to witness sentence of court martial in the case of the mutineers of the Texas schooner of war *San Antonio*, when the articles of war were read. The charges and specifications of charges also read against Frederick Shepperd (late Boatswain of the *San Antonio*) who was acquitted and released, J. McWilliams (seaman) who was not guilty of the 1st & 2nd charges but guilty of the 3rd, but recommended to mercy, and was pardoned and released from confinement. And William Barrington (seaman) who was guilty but in consideration of his information Mr. Dearborn (Lieutenant aboard the *SA*) at the last moment that a meeting was taking place, his sentence was 100 lashes with the Cats and told he would have it inflicted on the next day at Meridian.

The charges were Mutiny, Murder or an Attempt to Murder, and Desertion.

Latitude at Meridian 26°44'12" N

April 22ND

Fine breezes and the ship going alone finely, the brig not in company, and every 4 hours during the night fired a gun as a signal for her should she be in hearing. At 11 a.m. called all hands to witness punishment when the sentence of the court martial was executed on William Barrington.

Latitude observed at Meridian 25°36'20" N

April 23RD

This day pleasant weather. Ship jogging merrily for Telchac. Exercised the crew at general quarters.

Latitude observed at Meridian 26°26'45" N

April 24TH

Exercising our men. Ship going slowly along. Brig not in company.

Latitude at Meridian 25°35' N

April 25TH

This day beating for Telchac. At 11:30 a.m. called all hands to witness sentence of court martial in case of *San Antonio* when the articles of war were read and the charges and specifications of charges also read against Edward Keenan who was guilty of the 3rd charge and punished immediately and released. Antonio Landois (marine), William Simpson (Corporal Marines), Isaac Allen & James Hudgins (seamen) who were found guilty of all the charges and sentenced to be hung at the fore yardarm and given until Meridian next day to prepare to die. The crew was piped down and the prisoners were received on the quarter deck atop No. 9 gun.

APRIL 26TH

Still beating for Telchac. At 11:30 laid fore topsail to the mast and hoisted the colours. At 11:45 called all hands to execute sentence of court martial when they were addressed by Commodore Moore on the subject of Mutiny. At 12, the prisoners were carried forward and placed upon the scaffold. After addressing the crew, the ropes were placed around their necks. Until this time they appeared to believe they would be pardoned and did not evince much fear, but now the truth flashed upon them and they knew they had to pay the penalty of their crimes, and commenced praying eagerly and piteously for pardon. At 12:20, the signal gun was fired and the four prisoners run up to the fore yard.

Latitude at Meridian 23°31'03" N

APRIL 27TH

At 1:30 p.m. lowered the prisoners down and gave them to the master surgeon to prepare them for burial. At 1:40 filled away. At 2:30 laid the main topsail to the mast and called all hands to bury the dead, and after reading the funeral service over them, their earthly remains were committed to the deep. Filled away and stood on our course until 12 p.m. when we hove to. At 10:30 a.m. filled away and made sail. Brig not in company.

Latitude by observation 22°09'22" N

APRIL 28TH

Got everything ready for an engagement. At 6:45 hove to. At 4:30 made Telchac. Enemy's steamer *Montezuma* not there was expected. At 11:50 filled away and stood off and on. At daylight the *Wharton* in sight. At 4:15 laid mizzen topsail to the mast for her to come up with us.

FROM THE JOURNAL OF
MIDSHIPMAN JAMES L. MABRY,
1839-1840

The following excerpts are from a journal kept by Midshipman James L. Mabry, who arrived in Texas at the end of 1838 and entered the Texian navy in August of 1839 as a midshipman aboard the *San Antonio* (formerly the *Asp*) and later aboard the *Austin*. The excerpts presented here were originally published in the *Galveston Daily News* in three issues of that paper in January of 1893. The original manuscript journal was then in possession a Mrs. R. W. Shaw, a descendant of the Hurd navy men. Her father was Capt. James Hurd, formerly a lieutenant aboard the *Brutus,* and her grandfather was Capt. Norman Hurd, purser of the navy. In addition to Mabry's journal, she also possessed a badly deteriorated ration book and a naval account ledger dating from December of 1840. The selections presented here were at the choosing of the *News*.

Tuesday, October 23, 1839—These 24 hours, wind from the south and east. At 7 a.m. steamship *Columbia* drifted athwart our hawse. Rigged in jibboom and veered away ten fathoms on each chain to clear. At 7:30 she swung clear. Confined Charles Griffin, seaman, in double irons for disobedience of orders by order of Lieutenant G. G. Marion. At 11 a.m., the American brig *John Bartlett* drifted across our bows; veered in the end of both chains and cleared. Hove in forty-five fathoms of each chain.

Sunday, October 27—At 1 p.m. four men came on board from American schooner *Mary Eliza* (Capt. Higgins). At 5 p.m. one of the men from the *Mary Eliza* leaped overboard.

Sent a boat and picked him up. They all soon became mutinous and were secured in double irons at 10 p.m. They were sent on board their vessel.

Monday, November 4—Schooner *Louisville* came alongside and delivered 8 1/2 cords of wood and 750 gallons of water.

Saturday, November 9—Received in the master's department 1 barometer, 1 hanging compass, 1 thermometer, 1 sextant, 1 quadrant, 1 Blount's *Coast Pilot*, 1 nautical almanack of 1839 and one of 1840, 2 espying glasses, 1 chart of Gulf of Mexico, one 28-inch glass, two 14-inch glasses, 1 chronometer from the schooner *San Bernard*, 2 Bowdiche's *Practical Navigators*, 1 case of mathematical instruments. Received in gunner's department 10 patent carbines, 12 cap primers, 600 percussion caps and 12 levers. Received in boatswain's department 28-1/2 lbs marlin, 28-1/2 lbs houseline and 1/2 hank sewing twine. Received in purser's department 17 jackets, 34 flannel shirts and 17 pairs of brogan shoes.

Sunday, November 10—At 2 p.m. hoisted the jack and fired a gun for a pilot. Hove up the larboard and let go the starboard anchor, and vessel away thirty fathoms from chain. Received in purser's department for the officers 2 mattresses, 6 pillows, 24 sheets, 14 counterpanes, 12 pillow cases and 18 towels. Hove up starboard anchor and got underway and stood for the bar. At 9:30 stood back; at 10:30 let go anchor under the fort, which bore S.S.E.

Thursday, December 12—William Freer (gunner) and John Murphy and John Balock and James Fay (boys) deserted from the vessel with the gig at 11:30 a.m.

Monday, December 16—In the Mississippi River. At 6:30 John Smith, James Welch and E. Davis, all seamen, deserted from the vessel with the cutter. Having no other boat alongside we had to procure one from shore and pursue them. At 8 returned with the cutter.

Friday, January 3, 1840—Galveston Bay. Sent six men up the bay for food and water with twenty-one day's rations. Moored ship with an open hawse to the north and east.

Friday, January 10—At 10 a.m. information from Lieutenant Williamson of the desertion of John Wilson and John Loomis, who were temporarily transferred to the water tank on the Trinity River, who deserted on the 10th instant.

Saturday, January 11—This day moderate breezes from the westward and pleasant. Fired a salute of seventeen guns in honor of Mr. Henderson, minister plenipotentiary to England and France, who had just arrived.

Monday, January 20—Received in purser's department 16 barrels bread weighing 1,297 pounds, 2 barrels flour, 1 keg of butter weighing 70 pounds, 3 barrels beef, 8 barrels pork, 1 barrel vinegar of 25 gallons, 1 barrel of molasses of 41 gallons, 5 sacks of beans of 75 gallons, 7 boxes tea weighing 94-1/2 pounds, 4 boxes cheese weighing 157 pounds, 1 barrel rice of 90 gallons, and 1 barrel whiskey of 87 gallons.

Sunday, February 23—Schooner under command of Lieutenant O'Shaunessy. At 12 noon, ex-President Houston went on board the brig-of-war *Colorado*, Commander E. W. Moore, escorted by a number of citizens, and was saluted by seventeen guns. Received in purser's department 40 barrels of bread, 8 barrels flour, 11 barrels beef, 10 barrels pork, 8 barrels whiskey, 3 barrels rice, 3 barrels beans.

Tuesday, March 3—A national salute was fired by the fleet, in commemoration of the declaration of Texian Independence.

Saturday, March 7—At 10:15 a.m. the Texian steamer *Zavala* arrived from New Orleans and came to below the navy yard, after a passage of forty-eight hours.

Saturday, March 28—A seaman died on board the Texian steamer of war *Zavala*, and the flags of the different vessels of war were half-masted.

Sunday, March 29—A splendid ball was given on board the sloop-of-war *Austin* by the reefors of the fleet, said to excel anything of the kind that was ever given in Texas.

Thursday, April 23—Received from schooner *Striped Pig* six beakors water and one cask. Midshipman Burnett ordered by Captain Lothrop to take charge of the above schooner.

Friday, May 8—A marine shot at the navy yard in the act of desertion.

Sunday, May 10—The marine was buried at the navy yard. At midnight Jordan (seaman), [illegible] (captain's steward) and Johnson (landsman) deserted with the gig.

Monday, May 11—Lieutenant J. O'Shaunessy and Midshipman J. L. Mabry went in pursuit of those deserting men. Received one barrel water and thirty sticks wood from the *Striped Pig*.

Wednesday, June 3—Sloop-of-war *Austin*, bearing the broad pennant of Commodore E. W. Moore, President Lamar and suite came on board and were saluted with twenty-two guns.

Friday, June 5—Judge Lipscomb, secretary of state, came on board and was saluted with seventeen guns. Received for purser's department 200 tin pots and twelve quires paper and envelopes.

Tuesday, June 9—Received from navy yard one box boarding pikes, two boxes Roman swords, nine barrels whiskey, eighty gallons rice, three bags beans, 150 tin pans, two boxes [illegible] and one box cutlasses.

Tuesday, July 14—At 3 a.m. a man jumped overboard and was picked up by the cutter.

Thursday, July 16—At 9 a.m. the steamer *Columbia* ran into the schooner *San Jacinto* and carried away her anchor and part of the chain.

Thursday, July 30—At 6 a.m. made a strange sail on the lee bow and stood in chase. At 7:30 the strange sail hoisted the Mexican colors, distance three miles. At 11 hoisted United States flag and fired a gun. The strange schooner hove to. Sent a boat on board and brought off five large turtles. She proved to be the federal schooner *Picalina*, bound for New Orleans, loaded with logwood and turtles.

* * * * *

ON BOARD THE TEXAN WAR SHIP AUSTIN, FLYING THE BROAD PENNANT OF COMMODORE E. W. MOORE

August 20, 1840—Peak Orezaba, the highest mountain peak in North America and distant 120 miles. Bore south-west by west...At 11:30 discovered a strange sail on our weather bow. Made her out a schooner running along the land.

August 22—At 3:30, 1st and 2nd cutters returned from shore with fresh beef. At 5 sent 3rd cutter on shore for wood. At 5:30, *San Jacinto* made signals 6729, 4884 and 68. We made 223. At 6, strange sail proved to be a brig standing to the eastward...made all sail and tacked ship. At 8, strange sail on the weather bow, distant about 8 miles. At 12, strange sail on the weather bow, distant about 12 miles.

August 23—At 1, tacked ship, strange sail bearing S.S.E. At 1:30 hove to and communicated with *San Jacinto*. At 3:45 the chase visible from topsail yard.

August 25—Hove to a schooner going to Vera Cruz by firing a gun. Back on main topsail and sent second cutter alongside with papers. Filled away and hauled up N.W. by W. At 3:30 discovered a brig getting underway in port. Took in royals and hove to. Point Delgado bearing N.W. 3/4 N., Vera Cruz S. E. by E. At daylight made a sail on weather beam, standing down for us. At 5:30 a breeze sprang up from off the land; braced about and made for the strange sail; shook the reefs out of the topsails and set top gallant sails and jib. At 8:30 a boat came alongside from the brig *Penguin* which proved to be an English government vessel carrying six guns, with dispatches.

September 6—At 6:30 Rio Bravo del Norte bore W. 1/2 N., distant 2 miles. Mustered the crew and punished P. D. Fitzsimmons for desertion and Thomas Stewart for improper conduct.

September 19—At 1:30 set top gallant sails. At 2:30 crossed royal yard, wore ship, set mainsail, spanker, jib, flying jib, foretopmast, staysails and royals and stood in chase of strange ship who hoisted Spanish colors...at 5:30 strange ship tacked and stood for us. Beat to quarters and spoke her. She proved the Spanish corretto *Guerero*, mounting twenty-two guns.

October 1—At 5 made a vessel with a signal of distress, lying on the reef at the north end of the island (Labos). Sent life boat on shore to inquire if any of the inhabitants could pilot a boat out to her. At 6 the boat returned, unable to obtain any information or assistance. Sent life boat on shore to build a fire as a beacon to the vessel in distress. At 9, manned, provisioned and sent life boat and second cutter to the relief of the distressed vessel lying on the Banquilla reef. The second cutter returned, not being able to proceed against a heavy head sea.

October 5—At 8 o'clock a boat from the wreck with the captain and some of the passengers came alongside, also our life boat.

October 6—At 1:30 launch returned unable to go to the wreck. At 3:30 the life boat and second cutter returned, bringing the remainder of the crew, passengers and baggage.

October 7—First cutter returned from the wreck with passengers, baggage and a quantity of other articles.

Some of those rescued from the vessel on the Banquilla reef above referred to were Colonel Gomez and General Gomez and family. They were federalists, and were prisoners of the Mexican government. They were transferred from the Texan warship *Austin* to the schooner *Conchas* and taken to Vera Cruz for trial. The wrecked vessel above referred to was the Mexican brig *Segunda Fama*. The following letters, in connection with this wreck, are very interesting:

Tampico, October 14, 1840

Department of the Commandant General of Tamaulipas, To the commander of the Texas sloop-of-war *Austin:* The captain of the Mexican brig *Segunda Fama*, Pablo Alcedon, has placed in my hands your favor of yesterday, by which, and also by information received from him, I have heard of the terrible disaster which the passengers and crew of that vessel suffered near the Island of Labos, also the timely and efficient execution which you so successfully used in saving them from the wreck. With this motive as well as for the generous deportment which you observed toward the unfortunates of the *Segunda Fama*, while on board your vessel, as also for the kind and hospitable treatment which they

239

received, I return to you my most sincere thanks and assure you of my particular consideration and grateful acknowledgement.

God and liberty!
Joachim Rivas

Tampico, October 7, 1840
E. W. Moore, Esq.

Much Esteemed Sir:

Accident alone could have empowered me to offer this trifling homage to the kindness which you have been pleased to extend to me in the melancholy situation which fortune placed me, thus giving you an opportunity of manifesting your characteristic sentiments, assisting and bringing us from the island where a tempest had thrown us.

I take the liberty to beg your acceptance of the accompanying trifling present which circumstances prevent my extending as far as I should wish, and believe me, it is the result of feelings of gratitude which may it please heaven to give me an opportunity to show more fully. Please divide this with your officers and make to them my most sincere regards.

I send one demijohn of spirits, one loaf of sugar, some rice and beef. I beg that you will excuse the smallness of my offer, as it is absolutely impossible to send, as I wish, to a greater extent, or to send other things that would be useful on shipboard.

Goodbye, commodore. I wish you all happiness, and at all times and under all circumstances. Please command as you wish.

Your obedient servant,
Pablo Alcedan

Manifest of the cargo which was embarked in Vera Cruz on board the Mexican brig *Segunda Fama*, bound to Tampico, which vessel was lost at 12 o'clock on the night of October 3...near the Labos Island: 180 boxes vermicelli, 310 bags of flour, 319 bags of coffee, 133 bales of mats, 76 dozen grass baskets, 1 box containing two pictures of saints, 1 box sweetmeats, 3 bags split peas, 1 box books.

List of the crew and passengers of the *Segunda Fama*: Captain, Pablo Alcedan; mate, Jose I. Ramirez; boatswain, Jose Santos; cook, Cristobal Mesa; seamen, Manual Gregario, Ivan Arcedo, Ivan Molado, Alexander Meuh, Vincento Cruz, Pablo Valdez, Sebastian Diaz, and boy Manuel Garcia. Passengers: Don Joachim Gomez, wife, four children and four servants; Francisco Gomez, Ignacia Enriquez and son, and Augustin Arce.

The articles saved by the crew of the *Austin* were forty-nine sacks of flour, six sacks of coffee, sixteen bundles of mats, a lot of spars, boards, rigging, etc.

October 17—At 1:50 standing in for Tampico bar, and at 2:00 clewed up the topsails, hauled down the jib and came to with the larboard anchor. At 3:00 the pilotboat came alongside, and at 3:30 shoved off with the captain of the wrecked vessel (*Segunda Fama*). At 4:30 a launch from Tampico came alongside and at 5:30 shoved off with the wrecked passengers, crew and baggage. At 5:40 the wrecked brig's boat shoved off with some of the crew. Passed the messenger round the capstan.

October 18—A launch came alongside the ship with fruit. At 8:30 a sail hove in sight, standing down for the anchorage. At 4 she came to anchor a short distance ahead of us. She proved the English brig of war *Racer*.

October 19—A Mexican launch came alongside with fruit. Gave one sack and sixteen pounds of coffee in exchange.

October 20—At 4:30 the English brig of war *Racer* got underway and stood down toward us. Sent the life boat on board. She backed her main topsail. At 5 p.m. boat returned with English captain. At 5:30 he left the vessel.

October 21—At 2, the second cutter was fired at 3 times from the shore and very narrowly escaped destruction, the balls striking very close to her. We directed a gun at the fort and fired it but the distance was so great that it did not carry. At the same time hoisted a signal of recall to the second cutter. At 2:48 she returned.

October 23—At 2:30 James Garrett, second gunner, died of the scurvy. During the watch caught 2500 gallons of water.

October 27—At 2, tacked ship and brailed up the spanker. At 12, S. O. Sawyer fell from fore topgallant yard overboard and was lost.

November 4—At 1, sent first cutter with 228 gallons of water, one bag of coffee, two bags of flour and ten boxes of vermicelli to the schooner *San Jacinto* and the launch with two anchors and chain. The schooner was ashore, where she had been driven in a norther, having parted one of her anchors. At 6, sent the launch with the men to the *San Jacinto*. At 7, sent the first cutter to the *San Jacinto* with 217 gallons of water. The captain left the ship. At 7:30 the captain returned. At 10, the first cutter returned, bringing thirty gallons of lamp oil and one signal lantern, also a lot of wood.

November 14—Galveston harbor: At 8:30 the launch returned from the city with...4 boxes sugar, 5 bags fruit and vegetables, 1 box soap, 6 bags charcoal, a lot of grass, tar and lumber.

November 17—At 4, grounded on the bar in 9-1/2 feet of water. Carried out kedge and at 5:30 hauled over. At 6:30 came to in 4-1/2 fathoms of water with 25 fathoms of chain. At 9:30 sent second cutter down to the bar to weigh the kedge. At 11:40 got underway and proceeded up the river (one of the streams emptying into the Laguna de Tabasco, southern Mexico.) At 6:35, *San Bernard's* topsail caught in the limb of a tree and carried it away. The federal brig continued up the river. At 9 took the federal brig in tow.

November 21—Moving up the river in company with the squadron. At 3 the city of Tabasco hove in sight. Came to with larboard anchor in three fathoms water in front of the town. The rest of the squadron also came to. Edward Thornton, seaman, was secured...in consequence of using mutinous and abusive language to the officers, also attempting to incite the crew to mutiny.

November 23—A court martial convened aboard the *Zavala* for the trial of E. Thornton, charged with mutiny. At 11:30 General Anaya visited the ship.

November 27—At 3:30 Edward Thornton, prisoner, returned from the *Zavala*, also the witnesses.

November 29—During the morning watch a number of citizens visited the ship.

December 3—During the day the crew employed arranging for a ball to be given by Commodore E. W. Moore to the citizens of Tabasco. At 9 o'clock the ball commenced and was numerously attended. From 4 to 8 the ball broke up in consequence of rain...

December 6—The federal brig-of-war fired a salute of twenty-one guns. At 9:40 she got underway and hoisted the Texian ensign at the fore and fired a salute of seventeen guns. At 10 we answered it.

December 11—James Malcolm, landsman, died of the fever. At 10 the *Zavala* came alongside of and made fast to us.

December 13—At 6 called all hands to up anchor. Got underway and backed down the river with the *Zavala*. At 7 came to about three miles below the city.

December 14—Still moving down the river in company with the *Zavala*.

December 15—At 11:30 boarded and took in tow the Mexican schooner *Florentine*. At 1:30 came to anchor near the bank of the river for the *Zavala* to take in wood. At 2:30 boarded the Mexican schooner *Elizabeth* and brought her to under our stern.

December 16—At 3:30 got underway and cast off the two schooners, giving them permission to proceed up the river. At 6:30 came to anchor off the town of Frontera.

December 17—During the night James Duffries, ordinary seaman, died of fever.

December 18—Sent James Duffries, deceased, to be buried.

December 22—At 8 p.m. Samuel Edgerton, commodore's steward, died of yellow fever. At 10 a.m. sent his body ashore for interment.

December 25—Sent for Dr. Clarke of the *San Bernard* to visit the sick.

CRUISE REPORT:
WRECK OF THE SAN JACINTO

Owing to a strong southerly current and calm that night, the following day, the 11th of October, the vessel had drifted much to the south, and the wind hauling to the north and east we got in the bight to the leeward of Vera Cruz, and had to carry a press of sail to get out, having the land in sight to leeward. We had a very heavy head sea to contend with, which again endangered the foremast but the schooner made good weather of it, and behaved remarkably well. I cruised then, in the neighborhood of the Point, when hearing from Lieut. Moore that it was possible I might find you at the Islands off Tuspan, I made sail for them and returned immediately. Having had no opportunity to anchor on the coast, owing to the constant swell, and getting very short of water, I deemed it proper to come to his place. Accordingly on the morning of October 20th, I delivered to Lieut. Moore the dispatches for you and also a quantity of provisions, and parted company.

On the passage down here, we encountered a gale from the north and west, and hove to, the first and latter parts under a close reefed mainsail, the middle part under a balance reefed mainsail. During this gale, there were great fears entertained that the foremast would go. On the afternoon of the 29th October, made this Island but too late to get in. The day following made it again and hove to for a pilot from the schooner *San Bernard*, which vessel was lying here. At 2 p.m., was boarded by Midshipman Underhill who had been sent by Lieut. Williamson to give me the directions in.

I entered and came to near the *San Bernard*, letting go the anchor and kedge, having the larboard bow gun ready for letting go. When beating up for the entrance, the boat sent by the *San Bernard* was cut to the water's edge, capsized and two men who were in her slightly injured.

At 1:30 a.m. October 31st, the *San Jacinto* commenced dragging, having the anchor and kedge down, backed by one of the guns. At 1:50, her stern struck on the reef where she now lies. By 4, all the officers and men had left the vessel. At daylight, all who were able were set to work to save what was possible.

I am much indebted to Lieut. Williamson for his assistance, and force furnished. Nothing could have been done to have saved her. There was not room to get underway, and had she encountered another gale in the condition she was, the lives of all must have been placed in jeopardy by the loss of the foremast, which, in the opinion of the 1st and 2nd Lieuts, Boatswain and myself, would have happened.

Through the unremitting exertions of the officers and men, many things have been saved; in fact nearly everything will be saved by tomorrow evening, if one or two boats can be furnished. It may be considered presumption in an officer of my grade to comment on the merits of others, but I cannot close without recommending to your especial notice Lieuts. Gray and Oliver, and Boatswain Wills, for their untiring and unceasing exertions in endeavoring to save the vessel, and afterwards in saving public property. Accompanying this, I had you the Surgeon's Quarterly Report and the Monthly Muster Rolls.

I am &c,
(Signed) James O'Shaunessy

(MS, Navy Papers at Texas State Library)

About the Writings of
Lt. William A. Tennison

In writing his 1909 history of the Texian navies, Dr. Alexander Dienst drew heavily upon a collection of papers which he had purchased and labeled "Tennison's Journal and Papers". Dienst mentions, in a footnote on page 27 of his work, that he believed Tennison "selected some... articles in the current papers of that period" to include in his journal, as well as entries from the log books of Texian vessels other than those upon which Tennison served. This seemed a logical conclusion to draw since no man, not even a Texian sailor, could be on multiple ships at once. The "Tennison Papers" have stood for more than a century as the backbone for our knowledge of the doings of the navies of the Republic. But Dr. Dienst knew, contrary to his footnote, that what he was working with wasn't at all firsthand information penned by a witness to the events and committed to paper. In a note among his papers at the Dolph Briscoe Center for American History, he writes:

> These letters or manuscript copies are contemporaneous. Part of them are in William Tennyson's autograph. Mostly—I am sure— they are copied from articles in the *Telegraph and Texas Register* for years 1836-37-38. —Dr. Alex Dienst

He was right. A careful comparison of the original Tennison manuscript documents, now part of the Dienst Collection at the Center for American History, and digitized copies of newspapers of the day show that Tennison's writings, with the exception of maybe two documents, were copied from the *Telegraph & Texas Register* and various New Orleans and Galveston papers. In the case of some of the Galveston papers, no extant copies are to be found to cite the exact dates of the articles Tennison copied. But the language used in the entries, taken together with the fact Texian sailors spent a bulk of their time in Galveston, points directly to the papers of that city. In many instances, articles found in the *Telegraph* cite the Galveston papers from which they were drawn. With the exception of a word omitted or added here and there, the Tennison documents are direct copies of the published articles. The only two entries that seem suspiciously original are the first and last. In studying Tennison's writing elsewhere (documents and claims in the Texas State Library archives) it becomes apparent pretty quickly that he was not a man of many punctuation marks. The first folio of the Tennison papers is likewise unpunctuated, and the last is entirely too personal to have come from any other source. It lacks punctuation, as well, of course.

Dr. Dienst procured the papers for $20 from Mr. Ernest J. Cornibe, Sr. of Waco in 1900, at the recommendation of C. W. Raines. Mr. Cornibe may have acquired them from his father in 1872, as a notation in one of the documents indicates that his father served on board one of the Texian vessels. On the provenance of the documents, Mr. Cornibe was silent in his letters to Dienst. In one letter, Mr. Cornibe describes the collection as "the most complete manuscripts of the army and navy of Texas" and goes on to say that no other eyes have seen the papers other than those of Raines and one other, and that the collection was "the only one of the kind in existence." Indeed, Cornibe thought the Tennison papers to be, at least in some part, firsthand accounts of Texas naval experiences. That they are not does not completely diminish their value. A large body of the Republic era newspapers published at Galveston have perished though flood, fire and general degradation. The merit of Tennison's notes is in the preservation of news of the navy, though not the personal recounting of its history. They are transcribed here in their entirety, and any errors in transcription I take full credit for. The documents have been cited in every history of the Texian navies but, to my knowledge, have never been published in full. That is the purpose of their inclusion here.

William A. Tennison served in both navies and stuck around during the dead air between the Battle of Campeche and annexation to deliver the remains of the Texian fleet to the United States government. In between service in the two navies and in the lull from 1843-1846, Tennison would have had the time on his hands to jot down his naval time capsule. He obviously did it in two sittings. The items pertaining to the first navy are in ink in a steady hand. They aged well. The items pertaining to the second navy, although on the same period paper, are written in pencil. Time and friction were not as kind to the pencil jottings and apparently Mr. Cornibe's father, in an attempt to save the history he thought was singularly preserved in the papers, tried to trace over Tennison's pencilled writings as they faded. The result is an ugly mess, but still semilegible. A note written by Mr. Cornibe on a page Dienst labeled "Folio 535" indicates that the handwriting there, consistent with the handwriting covering Tennison's hand elsewhere, is his father's. It is written next to his father's name on a list of men who served aboard the *Austin*.

The Tennison writings are presented here in chronological order, although they do not appear as such in the original manuscript documents.

William A. Tennison was between sixteen and eighteen years old when he entered the Texian service. After officially transferring command of the Republic's vessels to the United States, he made for Washington, D.C. where he married Mary Virginia Brooke in September 1847. The couple had two daughters. Tennison joined the U. S. Navy and became a Third Lieutenant in the revenue service. In census records, he claims his place of birth as the District of Columbia, and as of the time of this writing, I can find no evidence to the contrary. In 1857, the United States Congress settled the case of Texian Navy officers and their claims to commissions in the U.S. Navy after annexation. The deal offered was five years' pay at the U. S. Officer's rate in lieu of a officer position in the navy. William Tennison was one of the Texian officers who availed them of that offer, and in 1858 is found purchasing two leagues of land on Clear Creek as part of a real estate development. By the time of the 1870 census, he has disappeared.

LT. WILLIAM A. TENNISON:
A TEXIAN SAILOR'S JOTTINGS

(This description of the career of the *Independence* is erroneous on several points. It slightly confuses the Battle of the Brazos and capture of the *Independence* with the final battle and wreck of the *Invincible*. It does not have the language of a newspaper article until the very end, nor does not have the language of a formal cruise report.)

The schooner of war *Independence* was bought by the Republic of Texas in the latter part of 1835 and was commanded by Captain Charles E. Hawkins, 1st Lieutenant W. Galeger, 2nd Lieutenant Mellis, Surgeon Leving, Purser Leveine, Midshipmen Wm A. Tennison, E. B. Harrington, Boatswain Robert Gyles, Gunner George Marion and 40 seamen; cruised from New Orleans and Galveston to Brazos River. Capt. Hawkins was promoted to the Command of the Navy. Lieutenant Geo. Wheelwright joined the *Independence* at the Brazos River. Wm. Brown, Captain Commanding, joined at the same time. After a short cruise down the coast of Mexican and making several captures and destroying all public property belonging to the enemy, such as small arms and ammunition—Commodore Hawkins always respects private property of the Mexicans—she returned to New Orleans to refit.

SECOND CRUISE

Officers attached to *Independence* March 20th, 1836

Commodore Charles E. Hawkins, Captain Geo. Wheelwright, 1st Lieut. Frank B. Wright, 2nd Lieut. J. W. Taylor, 3rd Lieut. J. T. K. Lothrop, Sailingmaster W. J. Bradburn, Lieut. Marines Thomas Crosby, Purser Wm P. Bradburn, Com'd Aide Wm. A. Tennison, Midshipmen E. B. Harrington and Joseph Hill, Boatswain Robert Gyles, Gunner Geo. Marion and 45 seamen.

About April 1836 set sail for the coast of Mexico where she captured 18 or 20 small vessels. In several of these there was considerable small arms and ammunition which were destroyed. Commodore Hawkins always respects private property. Sailed for the Island of Alacranes to wood and water, where she remained a week and recruited. Sail from there for the coast of Mexico, 18 hours from the Alacranes. Three vessels hove in sight; squared away and run down to them. Supposed them to be the merchant brigs under convoy, they proved to be the brigs of war *Urrea*, Capt. Machin; *Bravo*, Capt. Davis of *Boston*; [vessel name not entirely legible in MS but may well be the *Libertador* which was active at the time. Capt. Davis, at the capture of the *Independence* in 1837, was known to have commanded this Mexican vessel, while Capt. Thompson commanded the *Bravo*], Capt. Thom. Thompson. After making the discovery, hove to and prepared for action. Commodore Hawkins addressed the officers and crew. He told them to have confidence in him and that he felt confident that if he could not take one of the brigs, they could not take him. He then asked the officers and men if he should engage the enemy. It was received with 3 cheers to a man. After this, engaged the Mexicans, hauled off without doing much damage. Hove to all day expecting another attack. At sundown Commodore Hawkins [illegible at MS fold] he would be able to board one of the brigs, but failed. Finding it impossible to capture one of them, he sailed for the Brazos River where he expected to meet with the *Invincible* and *Brutus*, and with them sail for the coast of Mexico to engage the vessels of the enemy, but unfortunately for Commodore Hawkins and Texas, Capt. Brown of the *Invincible* and Capt. Wm. Hurd of the *Brutus* disobeyed Commodore Hawkins' order and, running to New York with the vessels, left the coast of Texas to the mercy of the enemy. Commodore Hawkins finding he could do nothing alone and very much in want of provision, sailed for New Orleans. Shortly after her arrival at New Orleans, he was taken sick with the small pox [illegible in MS].

Captain Geo. Wheelwright assumed command of the *Independence* after provisioning the vessel and getting her ready for sea. William H. Wharton, Minister from Texas to the United States was on his return home. Captain Wheelwright offered his vessel for his accommodation, so he took passage in her.

[NOTE: The Battle of Brazos River, in which the *Independence* was captured, took place up the Brazos near Velasco, in plain sight of the town and of the Secretary of the Navy. Tennison gives it, below, as taking place off Galveston and doesn't mention the fight up the river.

Officers attached to the *Independence* at the time of her capture

Geo. Wheelwright, Captain—John W. Taylor, 1st Lieutenant—J. T. K. Lothrop, 2nd Lieut.—Robert Cassin, 3rd Lieut.— W. P. Brannon, Purser—Leavy, Surgeon—Thomas Crosby, Lieut. Marines—Wm. A. Tennison, Capt's Aide—E. B. Harrington, Joseph Hill, Whitmore; Midshipmen— Robert Gyles, Boatswain, George Marion, Gunner—and 33 men. Passengers: Wm. H. Wharton, Capt. Darocher, [several names illegible in MS], George Estes.

Sailed about the 10th of April from New Orleans. On the 17th of April off Galveston, the *Independence* fell in with two Mexican brigs of war, Commodore Lopez. The *Libertado* and *Euterb*, and after a running engagement of 2-1/2 hours, ammunition all expended, most all of the standing riggings and running gear being cut away. Blowing heavy from the N & W, the brigs busy in their labors, and finding it impossible to escape, struck her flag to the enemy, rather it was cut away by a shot from the brig *Libertador*. Capt. Davis of *Boston* came on board and demanded a surrender. Capt. Geo. Wheelwright being badly wounded, John W. Taylor, the first lieutenant, was in command. On Davis's demand to surrender, Lt. Taylor told him "Sir, I am your prisoner, but my sword you shall never receive," and he threw it overboard. The *Independence* had several sick and wounded. Among the wounded was Capt. Wheelwright. The officers and crew of the *Independence* were divided between the 2 Mexican brigs and landed as prisoners at the Brazos Santiago and taken to Matamoros where they received harsh treatment for the first 3 months. But after that were well treated by the order of Bustamente, the President of Mexico. The officers of the *Independence* were ever grateful to Commodore Lopez for his kind treatment...[illegible in MS.]...After 8 months confinement in prison in Matamoros, they were set at liberty by President Bustamente. Wm. H. Wharton and Dr. Levy made their escape from prison by the consent of all the officers with the aid of Capt. Thom Thompson who was very kind to the prisoners at the time they were landed in Mexico. Any other information as regards the *Independence* will be thankfully received.

Writings of Wm. Tennison

(from the *Telegraph & Texas Register*, August 16, 1836)

The Texian schooner-of-war, *Invincible*, Capt. J. Brown, returned a few days ago from a cruise along the coast of Mexico, in company with the *Terrible*, Capt. Allen, without meeting a single Mexican armed vessel. The *Invincible* sent in a challenge to the Mexican brig-of-war *El Vencedor del Alamo* and the other vessels lying at Vera Cruz, which was not accepted on the pretext that the crews of the gallant challenged were not in a condition to fight for want of pay. The *Invincible's* crew landed at the river St. John & St. Paul, where they compelled Mexican inhabitants to bring them wood and water. Subsequently, the *Terrible* has sent into Galveston a prize, the Mexican sloop *Matilda*, loaded with dry goods and provisions, and taken between Sisal and Campeachy. It appears that the powerful Navy of Mexico, as well as their Grand Army, has caught the panic; and to get another fight, they must be sought for on *terra firma*, in their own country. The *Invincible* fell in with a British, a French and a Dutch vessel, all bound to Vera Cruz. The Frenchman was very much alarmed, for the stripes and single Star of our naval flag were to him an inexplicable mystery and he supposed it belonged to pirates. When informed that it was the flag of Texas, he was still more amazed: he had never heard of such a country, knew not where it was and, the creation of a new Republic not known to him and not even mentioned in the books he had read, appeared to him like a fairy tale. Captain Brown treated him politely and he went on his way rejoicing.

(excerpted from *Telegraph* of September 13, 1836,
likely first appeared in a Galveston newspaper)

Capt. J. M. Allen of the schooner-of-war *Terrible* left on a cruise on the 12th ult. The armed schooner *Independence* off the NE pass of the Mississippi bound on a cruise to Campeachy, and from there to Matamoros to wait for the war schooner *Invincible*, Capt. J. Brown, to blockade the port.

(*Telegraph*, September 13, 1836, "To The People of Texas")

The steamboat *Yellow Stone*, for Galveston Island...came to anchor about sun down of the same day, and Santa Anna with his suite was placed on board of the armed schooner *Independence*, under the command of Commodore Charles E. Hawkins, then lying at anchor in the harbor.

(*Telegraph*, October 11, 1836, from "Message of the President")

...The present condition of the navy is by no means commensurate with the importance of that arm of the public defence. The deficit of means has restrained the executive government from effecting any actual increase of its strength. Some efforts have been made to improve its organization. Conceiving it of importance that an immediate and responsible commander-in-chief should be created, who was himself a practical man and practically engaged in the service, I appointed, with the advice and consent of the cabinet, Charles E. Hawkins Esq., a gentleman whose gallantry and nautical science would grace any service, to that high office with the rank of Commodore. The operations of the navy have been as efficient as could have been expected. They have prevented any depredations on our coast, by the enemy, and have expelled his maritime forces from the gulf...

DAVID G. BURNET —PRESIDENT

(*Telegraph*, November 2, 1836, from "First Congress - First Session")

...Mr. Speaker: —Your committee on naval affairs, to whom was referred the report of the acting secretary of the navy, and a resolution calling for an increase of the same have instructed me to submit the following report, and the bill accompanying it. From the report of the acting secretary of the navy, that arm of the national defence appears to be in a most deplorable and crippled condition. The *Brutus* and *Invincible* are both in New York in a situation which prevents their services from being immediately available, and the *Liberty* is detained in New Orleans; thus, while momentarily in expectation of a blockade from the fleet of our enemy, our whole line of sea coast is defended by one national vessel, the *Independence*, mounting seven guns, and four small privateers, each pursuing its own prey and not immediately subject to the orders of this government. While our navy remains in this condition, it is in the power of the enemy at his pleasure to cut off our supplies, and to seize upon our sea ports.

It appears to your committee, that the error which has produced the present bad condition of our navy has been radical, and co-existent with its first formation: in order to raise a naval force sufficient to cope with that of the enemy, the government was forced to purchase such vessels as could be most easily procured; these were vessels either originally unfit for the purpose intended, or worn out in the merchant service. The consequence has been that these vessels demanded daily repairs and were seldom in a situation for active service. So far in our struggle with Mexico, our navy has proved adequate to the protection of our sea coast, and to the annoyance of the enemy—but the navy of the enemy has lately been increased by the addition of several vessels of the most splendid description. It therefore becomes imperiously necessary that our navy should be immediately increased in the same ratio. Your committee therefore suggest the immediate building or purchase of the following description of vessels:

One sloop of war, 600 tons, mounting 24 guns, probable cost $60,000.

One steam vessel, 359 tons, mounting 10 guns, probable cost $45,000.

Two schooners, 200 tons, mounting 11 guns, probable cost $30,000.

[NOTE: Tennison's manuscript recording of the congressional minutes read verbatim until this point. Here, instead of recording it as written, Tennison records what effectively shaped up to be the Second Texas Navy, indicating that he was likely writing after 1840, following the fitting out of the vessels. He even omits the steamer *Zavala*, which had been laid up in 1840, never to return to service. Here is Tennison's version:

One sloop of war, 600 tons, battery 24 guns, medium 24s
Two brigs, 18 guns each, medium 18s
Three schooners, 7 guns each, medium 9s]

(from the *Telegraph* November 19, 1836
quoting the *New Orleans Bulletin*)

The Texian privateer *Thomas Toby* (late *Dekalb*), Hoyt, commander, has been cruising off the ports of Vera Cruz, Sisal, Campeachy, Matamoros and Tampico, since the first week in October, and has captured, about the 12th inst., a Mexican schooner and sent her into Texas. She soon after run in towards the fort at the mouth of the river, and playing her "long tom" upon it for some time, without however, doing much damage, except frightening the good people of the town nearly out of their wits, who supposing her to be the van guard of the Texian Navy, turned out *en masse*, repaired to the fort and along the river banks, determined to repel any hostile movement of the imaginary Texian fleet. The commander of the privateer soon after transmitted a challenge to the commandant of Tampico, requesting a meeting with any armed Mexican vessel which might be in port; but, receiving no answer within a reasonable time, she stood off, and spoke [to] the *Louisiana*, determined to capture all Mexican property she fell in with...

(from the *Telegraph* April 4, 1837)

Midshipman Waite, of the *Invincible*, tossed two bullets into the window of a Mr. Thomas, in New York. This has been magnified by the *N. Y. Sun* into, not a fish story, nor a moon story, but a love story.

(from the *Telegraph* May 2, 1837)

We stop the press to state that the sloop of war *Boston* is now at anchor off Galveston. The lieutenant of this vessel hailed Captain Hurd of the *Brutus* and stated that the *Independence* had just returned to the Mississippi just before the *Boston* left. Capt. Hurd has just arrived and informs us

259

that the search of the *Brutus* and *Tom Toby* has been un-
successful, no Mexican vessels are on the coast. Our whole
fleet will probably start for the Mexican coast immediately
after obtaining the necessary supplies of provisions.

(*Telegraph*, May 9, 1837, from "President's Message")

...The insufficiency of our navy must be the subject of serious consideration. When the constitutional government
assumed its functions, the armed vessels *Brutus* and *Invincible* were in the port of New York, and remained there
until a few weeks past, when they returned, but without
either crews or provisions for a cruise.

The *Independence* having not more than two weeks
provisions was taken to New Orleans some months since,
where she has been detained, and has not yet reported to
this government for service.

At an early day, a confidential officer was despatched to
the United States for the purpose of purchasing such vessels as would enable us to keep command of the gulf from
our enemy.

He has reported to the proper department, and his arrival is daily expected with one or more fine vessels, in
preparation to defend our commerce, and make reprisals
on the enemy....

—SAM HOUSTON.

(*Telegraph*, June 8, 1837)

The *Tom Toby* has just sent into Galveston harbour a
very valuable prize, being a large fine brig, strongly built
and capable [of] being fitted out as a man-of-war, bearing guns heavier than any now in the Mexican squadron.
She was captured on the coast of Campeachy and has on
board 200 tons of salt. The *Tom Toby* when last seen was
in hot pursuit of two Mexican schooners, this pursuit will

undoubtedly prove successful as "Fortune ever favors the brave." It is gratifying to reflect that our flag flaunts over one brave band, whose dauntless spirits delight to career with the "Stormy petrel" over the tossing billows where danger lights the "Path to glory and fame."

(*Telegraph*, June 8, 1837)

We rejoice that we are at length enabled to furnish the official account of the capture of the *Independence*. We have hitherto foreborne offering any comments upon the former vague accounts of this transaction, as we felt confident that many important facts had been overlooked which would completely exculpate our gallant tars from any disparaging imputations. We confess that when the first news of this combat arrived, containing the intelligence that the *Independence* had surrendered to two Mexican brigs without having received any injury, and her crew unhurt, a flash of shame and indignation mantled on our cheeks and the exclamation "30 or 40 cowards and an old hulk are no loss," almost involuntarily fell from our lips; better we thought it would have been if this crew dauntlessly nailing their unsullied flag to the mast head, hurling their mortal defiance to the grovelling foe—had fought on, and on, shouting the stern war cry of "victory or death," until the star of Texas, like the "star of day" went down in glory beneath the blood red billows, whose foaming crests were ringing to the last exulting cry of the *unconquered band of freemen!*

But the following statements have fully convinced us that we did injustice to those gallant tars in harboring even for a moment a thought so unworthy of them and of the Texian name.

Far from blaming them for this surrender, we rejoice that they may yet be preserved to ride through the battle storm which shall rend the tyrant banner from the mast it disgraces. This desperate and protracted conflict will long hold a prominent place in the annals of Texas and, like the fall of the Alamo, it shall inspire our children with ennobling sentiments. No flash of shame shall redden their youthful cheeks as they read the page which declares that thirty-one Texans, six only of these seamen, in a slow sailing armed schooner mounting only six *sixes* and one *long nine* fought four hours and a half, two Mexican armed brigs, one mounting "16 medium eighteens" with a crew of 140 men; the other mounting "8 brass 12 pounders" and one long eighteen midship, with a crew of 120 men!! One is astonished in reflecting that this little vessel was not annihilated by the first broadside from her powerful opponents, her dauntless little crew appear to have been preserved almost by a miracle, and it is cheering to reflect that their heroic conduct has furnished new proofs that our national escutcheon yet remains bright and untarnished. True the flag of our country has *once* been struck on the stormy billows of the Gulf, but like the Roman eagle stooping before the sword of Epirus, it has wrung from the abashed conqueror the bitter confession "Such men are invincible."

BRAZOS DE ST. IAGO, April 21, 1837

To the honorable S. RHOADS FISHER,
Secretary of the Navy.

SIR.—I have the honor hereby to transmit you an account of the late engagement between our government vessel *Independence* and two of the enemy's brigs of war, one the *Libertador* of sixteen eighteen-pounders, 140; the other, the *Vincedor del Alamo*, mounting six twelve-pounders, and a long eighteen a-midships, with one hundred men. Captain Wheelwright having during the action received a very dangerous wound, the duty of sending this melancholy communication has devolved upon me. To wit:

On the morning of the 17th, in latitude 29 deg. N., longitude 95 deg. 20 min. W, at 5 h. 30 m. A.M. discovered two sails about 6 miles to windward; immediately beat to quarters; upon making us out they bore down for us with all sail set, signalized, and then spoke each other. At 9 h. 30 m., the *Vincedor del Alamo* bore away, getting in our wake to rake us, the *Libertador* keeping well on our weather quarter, we immediately hoisted our colors at the peak. The enemy in a few minutes hoisting theirs, the *Libertador* on our weather quarter edging down for us all the time, till within about one mile, gave us a broadside, without wounding any of our men or doing other damage. The fire was at the same time returned from our weather battery, consisting of three sixes and a pivot, a long nine, the wind blowing fresh, and from our extreme lowness our lee guns were continually under water, and even the weather ones occasionally dipped their muzzles quite under.

The firing on both sides was thus briskly kept up for nearly two hours, the raking shots from the *Vincedor* in our wake, nearly all passing over our heads, as yet sustaining but trifling injury. At 9 h. 30 m. the *Libertador* on our weather quarter bore away and run down till within two cable lengths of us, luffed and gave us a broadside of round shot, grape and canister, while all this time the brig *Vincedor* in our wake continued her raking fire. Notwithstanding this we still continued on our course for Velasco, maintaining a hot action for full 15 minutes, with some effect upon her sails and rigging. The *Libertador* now hauled her wind, widening her distance, apparently wishing to be further from us, when she again opened her fire, which was on our part kept up without cessation. At 11 A.M. she bore away, run down close to our quarter and gave us another broadside of round shot, grape and canister, which told plainly on our sails and rigging. As before she again hauled her wind to her former position, and played us briskly with round shot, one of which struck our hull going through our copper and buried itself in her side.

At 11 h. 30 m. A.M., a round shot passed through our quarter gallery, against which Captain Wheelwright was leaning, inflicted a severe wound on his right side, knocked the speaking trumpet out of his hand, terribly lacerating three of his fingers. He was conveyed below to the surgeon, leaving orders with me to continue the action. We still held on our course in our respective positions, keeping up an incessant fire, for a full half hour, when the enemy signalized; then the *Vincedor* in our wake luffed up and gained well on our weather quarter. At that time, the *Libertador*, on our weather beam, bore away and ran down under our stern within pistol shot, our decks being completely exposed to her whole broadside, and at the same time open to the raking fire of the *Vincedor* on our weather quarter. In this situation, further resistance being utterly fruitless, and our attempts to beach the vessel ineffectual, I received orders, form Captain Wheelwright, to surrender, which was done.

The only damage done to our vessel, was that of parting some of our rigging, splitting the sails, a round shot in her hull, and the quarter gallery, which was shot away. Captain Wheelwright was the only person wounded on board. We shot away the *Libertador's* main top gallant mast, unshipped one of her gun carriages, took a chip off the after part of her foremast, killed two men, and cut her sails and rigging severely. We were immediately boarded by Captain Davis of the *Libertador*, who pledged his honor and that of Commodore Lopez, who was then on board, that we should receive honorable treatment as prisoners of war, as officers and gentlemen, and as soon as an exchange could be effected, we should be sent home. The kind attention and courtesy we have received from Commodore Lopez, Captain Davis and officers has been truly great, for which we tender them our sincere thanks, likewise Captain Thompson of the schooner of war *Bravo* has extended every civility and kindness. We leave this place tomorrow for Matamoros: what disposition will be made of us I know not.

Besides the officers and crew of our vessel, we had on board as passengers the honorable Wm. H. Wharton; Mr. Levy, Surgeon T. N.; Captain Darocher, T. A.; Mr. Thayer of Boston; Mr. Wooster, English subject; George Estess, acting lieutenant, T. N.; and Mr. Henry Childs.

I remain very respectfully, your obedient servant,

J. W. TAYLOR, Lieut.

P.S. Our crew consisted of 31 men and boys, besides the officers, out of this number there were six seamen, the balance not knowing one part of the ship from the other, and it was with great difficulty that we obtained this crew while in New Orleans.

(brief inventory of Mexican navy
excerpted from *Telegraph*, June 8, 1837)

Mexican Navy— (1) *Libertador* (2) *Iturbide*, 16 guns each (3) Brig *Vincedor del Alamo*, 8 guns, 12s, brass (4) Brig *Teran* (5) Schooner *Bravo*, one 12, double fortified bell metal, by Thomas Thompson

(*Telegraph*, August 22, 1837, taken from *Matagorda Bulletin*)

By the arrival of Midshipman Robert Foster, in charge of the Mexican schooner *Alispa*, a prize of the Texian schooner of war *Invincible*, we learn that our navy has been engaged for the last three months to some purpose. On leaving Galveston in May last, they proceeded to the mouth of the Mississippi, and cruised thereabouts seven or eight days without meeting anything of the desired description. They then altered their course for the coast of Mexico, and at Mugere's Island fell in with several pirogues, or schooner canoes; these were generally laden with articles of little value, and, with one exception, they obtained from them only their sails and provisions, with which they were able to keep themselves supplied, having originally laid in only a two months stock. In one, however, they found a cargo of logwood, which the captain of it redeemed with 600 dollars, on arriving at Sisal. This place they cannonaded about three hours, when owing to their destitute condition in the way of rigging, they thought it prudent to haul off. Had they have had spare rigging, with which they could have repaired in case of sustaining damage, they would have taken the town.

The *Alispa* is of about eighty tons burthen, and laden with crockery and hardware.

Another schooner, the *Telegraph*, captured by the *Brutus*, is expected at this port hourly.

An English brig, the *Eliza Russell*, of 180 tons, chartered by a Mexican house, laden with a general assortment of merchandise, was taken off Alacranes, by the *Brutus*, and has been sent to Galveston.

Our men made repeated landings, and on the cruise burned to the ground eight or nine towns, some of them of considerable size.

The Mexican fleet is lying at Vera Cruz, unmanned.

From Midshipman Foster, we learn that our townsman, the Secretary of the Navy, is in fine health and spirits, and may be expected to return with our vessels in a few days. He and captain Boylan, of the *Brutus*, on one occasion made rather a narrow escape. Landing at one of the towns before mentioned, with a boat's crew of six, they hauled their boats upon the beach and moved in a body two or three hundred yards from it, without their arms. At this juncture, a body of cavalry of fifteen or twenty wheeled upon them, and they had barely time to recover the boat in safety. Judge Fisher, having a pistol by him, made use of it, dropping one of the assailants.

At another time, the Boatswain of the *Brutus*, with four men, in pursuit of a pirogue, was benighted; and landing at a small village, he laid it under contribution, rousing the Alcalde from his sleep for the purpose. He raised fifty dollars, which is said to have been tight work, and left early the next morning.

(*Telegraph*, September 2, 1837)

We are informed by a young man who left Galveston on the 29th ult. that the *Invincible* and *Brutus* arrived off Galveston on Saturday last, having a Mexican armed schooner in tow, which they captured near the banks of Campeachy. The *Brutus* and this prize entered the harbour in safety on the same evening, but the wind proving unfavorable for the entrance of the *Invincible*, she lay to off the coast until morning. But when morning came, she was discovered to be between two Mexican armed brigs, one to the windward and the other to the leeward. The *Brutus* immediately prepared for action, and got under way to assist her, but from some untoward accident ran aground before reaching the bar and was compelled to wait until next day for return tide, in order to get off.

The *Invincible* in the meantime nobly sustained herself in the unequal encounter. Throwing abroad the one-starred flag to the freshening breeze, she bore down upon the commodore's brig and a desperate engagement ensued which lasted until sunset, both brigs keeping up an ineffectual fire upon her during the whole day. Not a man on board was killed and she received hardly any injury; the brigs however were both seriously injured. The commodore's vessel received several shots in her hull, had her main gaft carried away, a large portion of her bulwarks stove in and her flag shot down; this latter occurrence induced the spectators on shore to believe that Commodore Lopez had struck to this little schooner. Both brigs could easily outsail her, yet the enemy did not dare to board during the whole engagement, and were forced repeatedly by the desperate bravery of the *Invincible's* crew to haul off from close combat.

Towards evening, the *Invincible* relinquishing the contest attempted to enter the harbour, but the water being still quite shallow on the bar and the wind unfavorable, she struck on the breakers near the southeast channel. The crew all landed in safety and night closing in hid her from view. In the morning she had entirely disappeared, having gone to pieces during the night! Thus has perished the most noted and most successful, although the poorest vessel [NOTE: Tennison omits "the poorest" in his transcription of the article] of the Texian navy. She has fallen like the strong man in his prime, and many a brave tar will lament for her as for a dear friend. Farewell then to thee, gallant bark, true to the last to they proud name, thy foaming path has glowed with brilliant exploits, and victory, amid the tumults of every battle storm, has hailed thee triumphant! Invincible to the last, the red bolts of war have hurtled around thee in vain, with thy flag flying to the peak, with thy daring crew in safety by thy side, shouting defiance to the foe, thou hast gone down in glory, amid those sounding billows which so often witnessed thy prowess, and run to they conquering thunders.

(*Telegraph*, September 9, 1837)

PUBLIC MEETING

At a large and respectable meeting of the citizens of Houston, assembled on the 4th inst., for the purpose of expressing their high admiration of the character of the honorable S. Rhoads Fisher, secretary of the navy of Texas, and of tendering him an invitation to partake of a public dinner, Dr. M. Forrest was called to the chair, and Mr. J. W. Eldridge was chosen secretary.

On motion it was

Resolved, That we hail with joy the safe return of the honorable S. Rhoads Fisher to the seat of government, and that, in order to express publicly the regard in which we hold him, as an officer and a gentleman, he be invited to partake of a public dinner.

Resolved, That a committee of eight persons be appointed by the chair, to wait upon Judge Fisher, and tender him in behalf of the citizens, an invitation to a public dinner, and to make all necessary arrangements for the provision of the same.

The following gentlemen were then appointed a committee to carry into effect the object of the foregoing resolution, viz: Messrs J. Birdsall, T. J. Gazley, H. McLeod, W. L. Cazeneau, N. F. Smith, F. Moore, Jr., L. Bicknell and A. Henriques.

Resolved, That the proceedings of this meeting, and the correspondence be published in the *Telegraph and Texas Register*.

Resolved, That the meeting adjourn.

[Signed.]   MOREAU FORREST, *chairman.*
                   J. W. ELDRIDGE, *secretary.*

CORRESPONDENCE

Houston, September 4th, 1837

To the honorable S. Rhoads Fisher,
secretary of the navy of Texas.

Sir:—The undersigned have been appointed to a committee of the citizens of Houston, to greet your safe return among us, and to offer you a public dinner as a testimonial of their regard for the able and energetic manner in which you have conducted your department, thereby enabling our little navy to extend the sphere of its operations to the very homes of our enemy. Its gallant exploits under your own eye have thrown a halo around the "single star" which will animate the future hero, and be the glorious beacon of light of coming combat.

We are happy, sir, in being the organ of our fellow citizens upon this interesting occasion, and with high regard for your worth and admiration of your character.

We have the honor to be, dear sir,
Your most obedient servants.
JOHN BIRDSALL,
T. J. GAZLEY,
H. MCLEOD,
W. L. CAZENEAU,
L. BICKNELL,
N. F. SMITH,
F. MOORE, JR.,
A. HENRIQUES.

CITY OF HOUSTON, Sep. 4, 1837

To Messrs. J. Birdsall, T. J. Gazley, H. McLeod, W. L. Cazeneau, L. Bicknell, N. F. Smith, F. Moore, Jun., and Alex Henriques, committee on behalf of the citizens of Houston

FRIENDS AND FELLOW CITIZENS:—It is with no ordinary feelings of heartfelt gratification that I acknowledge to have received your polite note of this day, tendering to me on behalf of the citizens of this patriotic and flourishing city a public dinner as expressive of their approbation of the course which the head of the Navy Department has recently pursued. I thank you gentlemen, sincerely thank you, not only because it is highly gratifying to my personal feelings to have my official conduct approved by the PEOPLE but because I feel myself so identified with the officers and crews of our gallant little navy, that I consider any compliment paid to the head of the department as extended to them. The "gallant exploits" achieved "under my own eye," are attributed to the skill, courage and determination of our officers and crews, and if our Congress will only extend its fostering protection and support to our navy, the names of Wheelwright, Thompson and Boylan will stand brightly conspicuous in the pages of our nation's history.

My motives for taking passage on board our vessels of war on their recent cruise will be explained by presenting to you an extract of a letter dated June 17th addressed to a friend and correspondent in New York, (Dr. Bartlett, editor of the *Albion,)* and is in these words:

"Schooner *Invincible*
Off Passo Caballo, June 17, 1837

It is ten days since I left Houston and immediately joined our little squadron, then lying in Galveston Bay, and after convoying the schooner *Texas*, laden with government stores to Matagorda Bay, up helm and bare away for

Galveston, to receive orders from the president; we shall
be there tomorrow, and shall stretch to the southward
with the hope of falling in with the enemy. I am a volunteer, I cannot precisely say amateur; but I have thought
for some time upon the expediency of personally taking a
part with the navy, and have decided it was right. I know,
you gentlemen of systematized governments will smile
at the idea of the "secretary of the navy" turning sailor,
and may be inclined to consider it better adapted to the
adventure seeking disposition of the knight of the rueful
countenance; but my opinion is that it will inspire great
confidence in the men, and stimulate our congress to do
something for us; for it appears that this branch of the national defence has never been popular in its infancy in any
country: it ever has been compelled to fight itself into notice and government patronage; such at least I am satisfied
is our case, and I think my present step is precisely such
as will suit the meridian of the views of our Texas population. We must be governed and actuated by such course as
may best suit *us*: we are acting and legislating for ourselves
and not for the world, and however at variance our system
of policy may be with the preconceived ideas of right or
wrong amongst the world at large, I humbly conceive that
as we have to lie in the bed, we have the right to make it.
Therefore it is that however quixotic my present step may
appear, and indeed for the United States or Great Britain
would be, *I* am satisfied it is right."

Having now, gentlemen, as the representatives of the citizens of Houston, shown you some of my motives for my
recent course, and expressed to you my gratified feelings
for the confidence reposed in me, as an officer and man,
it becomes my duty frankly to state, and with equal frank-
ness to regret, that circumstances of a peculiar nature will
prevent my accepting of the proposed compliment and
meeting you at the social board where the "feast of reason

and the flow of soul," while it would warm us into an increased love of country would endear us to each other. But we have met—and will meet again. In our private relations festivity shall unite us and love of Texas make us a band of *Brothers*.

To you Gentlemen of the committee, permit me to say how truly indebted I feel, for the handsome manner in which you have expressed the sentiments of my fellow citizens generally; like the diamond, its beauty, its lustre, its value is shown by the ingenuity of the artist.

And in now, taking my leave, believe me when I say, that whatever situation the people may call me, I shall feel honored by uniting my efforts to theirs.

God prosper Texas! and to you my friends and fellow citizens, happiness and thanks.

S. RHOADS FISHER.

(Source unknown, though an account was published in the October 11, 1837 *Telegraph*. Racer's storm actually raged from the 6th-8th)

Brutus and *Tom Toby* both lost on the 10th of October 1837 in Galveston Harbor during a heavy gale.

[EDITOR'S NOTE: Secretary Fisher, reporting on September 4, 1837 for duty in Houston, after his extended cruise with the navy, was soon after charged by Sam Houston with many accusations and suspended from office. Fisher's comments in the above *Telegraph* letter were among the many things that had President Houston hot under the collar. The entry which follows is one of many defenses of Secretary Fisher after it was made public that he was to be tried for his character while on an extended cruise without the "approbation of the President."]

(Not found in the *Telegraph*. Dated April 17, 1837 in Tennison's papers, but probably published in October or November of that year. Likely a Houston newspaper defense of Secretary Fisher, who had been suspended from office by Sam Houston in October 1837 for, among other charges, taking leave of his official duties to take a three month cruise aboard the *Invincible* as a volunteer sailor, actions taken during that cruise and "offensive" remarks in the *Telegraph* upon his return to Houston on September 4, 1837. Newspapers, citizens, naval officers, David G. Burnet and John A. Wharton sprang to Fisher's defense.)

On the disastrous morning of the 17th of April we were present at Velasco where the *Independence*, struggling like a tiger in the toils, hove in sight, closely pursued by two Mexican brigs. Large crowds of citizens had collected upon the beach who were curiously watching the progress of the combat, and occasionally given vent to their excited feelings, by casting in no measured terms, scoffs, jeers, and imprecations upon that "imbecile government," which had permitted our navy to lay snugly moored in the harbor of the United States when they might with united force have been enabled, not only to have banished from the Gulf every hostile armament but, have kept the whole Mexican coast in a state of continual alarm. S. R. Fisher was present and evidently listened to these remarks with the deepest mortification and chagrin. He appeared to feel that they were all intended directly for him. In conversation with him upon the subject, he acquainted us with his full determination to effect something to retrieve the lost character of the navy.

Upon his arrival at Houston, he immediately made every possible exertion to furnish the necessary means of fitting out the *Brutus* and *Invincible* for a cruise. We witnessed these exertions with deep emotion, for we felt that the very honor of the nation was in some degree connected with them. The lapse of a few weeks found these two vessels upon the gulf, half manned [and] half rigged, with old worn out sails and a scanty supply of miserable provisions, for a six weeks' cruise, but all these deficiencies were more than counterbalanced by the determined spirit of the little crews and their heroic commanders. The secretary of the navy was a *passenger* on board by virtue of furlough from the President; he nobly preferred encountering the perils and privation of a dangerous cruise to vaporing away the time in whiffs of tobacco at Houston.

They proceeded to the coast of the enemy and fearlessly offered combat to a force ten times greater than their own; with a courage unequalled in that quarter since the days of the Buckaneer, they boldly sailed into the port of Sisal and cannonaded the fort and town during the space of four hours, at the end of which time they retired uninjured. They captured six valuable prizes and, during the space of nearly three months, they kept the whole Mexican coast from the Brazos Santiago to the extremity of the Yucatan in a state of continual alarm.

They returned flushed with success to the port of Galveston and there, as if fortune had been desirous of crowning this cruise with a blaze of glory, the little *Invincible* singly encountered and actually beat off two Mexican brigs, either of which was of more than ten times her force. Retiring victorious from the contest, she was lost on the verge of the harbor, but her gallant crew and lionhearted commander landed in safety, stript of everything but their rags and that glory which is imperishable. There was no hope which supported them, it was the proved reflection that *they had done their duty*, and that the gratitude of an admiring nation would dictate their reward, and they looked forward with pleasure to the period when a *just* and beneficent government would extend to them its fostering hand with ardent joy.

(*Telegraph*, November 11, 1837)

A MIRACLE! A MIRACLE!—The Texians who were imprisoned at Matamoros have all been released by the order of President Bustamente. Lieutenant Taylor, Lothrop and their comrades arrived in Galveston Bay on the 4th inst. They had been sent round in the *Louisiana*, which was ch[artered by the Mexican government for this purpose!]

(from the *Telegraph's* list of events that occurred in the second year of Texas independence, published March 17, 1838)

March 3, 1837 - The resolution of the congress of the U.S.A. acknowledging the independence of Texas is signed by President Jackson.

March 17 - The skirmish near Laredo, in which Deaf Smith with twenty men defeated forty Mexican cavalry, kills ten of their numbers and captures twenty horses.

August 27 - The *Brutus* and *Invincible* after a cruise of nearly 3 months on the coast of Mexico, during which they captured 6 prizes, arrived in Galveston Bay.

November 4 - The captive Texians of the *Independence* who have been released by order of President Bustamente arrived at Galveston.

November 30 - Deaf Smith, the celebrated spy of Texas, dies at Fort Bend.

(likely taken from a Galveston paper, as the verbiage indicates)

June 24, 1840

The following is a catalog of the vessels and officers which leave port today under sealed orders. We hope shortly to chronicle movements on their part,—interesting to the public and honorable to our young republic.

LIST OF OFFICERS ATTACHED
TO THE SLOOP-OF-WAR *AUSTIN*

Commodore Edwin W. Moore, Commanding
1st Lieutenant - E. P. Kennedy
2nd Lieut. - D. H. Crisp
3rd Lieut. - J. H. Baker
4th Lieut. - Wm. Seeger
Acting Master - C. Cumming
Surgeon - J. B. Gardner
Purser - N. Hurd
Lieut. of Marines - T. W. Sweet
Commodore's Secretary - C. A. Cristman

Midshipmen:
C. B. Snow
G. F. Fuller
M. H. Dearborne
L. E. Bennett
J. C. Bronough
E. A. Wezman
W. W. McFarlane

Boatswain - John W. Brown
Gunner - Joseph Salter
Carpenter - Wm. Smith
Sailmaker - C. Cremer

Steam Ship *Zavala*, 8 guns

Captain - J. T. K. Lothrop
1st Lieut. - George Henderson
2nd Lieut. - W. C. Brashear
Master - Daniel Lloyd
Surgeon - T. P. Anderson
Purser - W. T. Maury
Capt. of Marines - J. W. C. Parker
Chief Engineer - G. Beatty
Capt's Clerk - R. Bach

Midshipmen:
C. Betts
C. C. Cox
J. E. Barrow
H. S. Garlick
J. A. Hartman

Boatswain - James Crout
Gunner - T. Howard
Carpenter - Joseph Auld

Schooner *San Bernard*, 5 guns

Lieut. Commanding - W. S. Williamson
1st Lieut. - George W. Estes
2nd Lieut. - Wm. A. Tennison
Surgeons- Charles B. Snow & R. M. Clark
Purser - A. G. Stephens
Capt's Clerk - W. H. Brewster

Midshipmen:
C. B. Underhill
John P. Stoneall

J. B. F. Bernard
L. H. Smith

Boatswain - George Brown

SCHOONER *SAN JACINTO*, 5 GUNS

Lieut. Commanding - W. R. Postell
1st Lieut. - J. O'Shaunessy
2nd Lieut. - A. G. Gray
Acting Master - Wm. Oliver
Surgeon - Fletcher Dory
Purser - Robert Oliver
Capt's Clerk - J. J. Tucker

Midshipmen:
C. S. Arcamble
A. Walke
J. O. Parker
W. T. Bell

Boatswain - G. W. Wills
[illegible] - E. Horton

SCHOONER *SAN ANTONIO*, 5 GUNS

Lieut. Commanding - Alex Moore
1st Lieut. - Thomas Wood, Jr.
2nd Lieut. - A. J. Lewis
Acting Master - A. A. Waite
Purser - James W. Moore
Capt's Clerk - Hugh A. Goldsborough

Midshipmen:
James A. Wheeler
E. F. Wells
L. M. Minor

Boatswain - Hugh Schofield

(*New Orleans Commercial Bulletin* September 21, 1840.)

FROM MEXICO

Our Havana papers contain extracts from Mexican ga-
zettes from which we make the following extracts:

Vera Cruz, August 22, 1840

There are three Texian vessels of war laying off in our
port, at the distance of a few leagues. We are, in a manner,
blockaded by the Texians; and although pirates, will not
be permitted, with impunity, to molest vessels of other nations. This, we confess, makes our blood boil. We cannot
view it with serenity; it makes us desperate. And in the
midst of a useless wrath, we are compelled to acknowledge
the cause of our present humiliating situation. A banditti
insult of a nation has this day the elements to place it in
the foremost rank of those discovered by Christopher Co-
lumbus. Who, on such reflection, is not saddened, is not
enraged, or who is not warmed with the fire of patriotism
to punish such temerity?

Great as are the difficulties in the way of revenging this
insult, they are not insuperable. We have no marine, nor
are we likely to have one. And what, then, is the remedy?
Justice and energy, and decision to chastise the criminals,
to organize the army and assert our rights on the Banks of
the Sabine. Without this, we shall never have peace. For

our present apparent supineness only emboldens the insurgent Texians to give us fresh insults.

(likely from a letter written by Purser Maury, a.k.a. Murray, of the *Zavala* to the *Galveston Courier* from which the *Telegraph* of December 23, 1840 quotes by way of the *Morning Star*. As of this writing, we have not been able to locate an extant copy of the *Courier*)

Texan Steam Ship of War, *Zavala*
Town of Frontera, River Tabasco

November 7, 1840

We left New Orleans on the 18th July last, and proceeded to the Arcas Islands, and after cruising about a little we had to go to Sisal (a town on the coast of Yucatan) for wood, on the 15th August; where we were detained until 23rd September, when we went to Arcas where we expected to meet the Commodore and obtain a supply of provisions from him—but unfortunately he was not there, and after waiting a week on half allowance we went to Laguna de Terminos to obtain provisions. We got enough provisions there by giving draft on the Consul in New Orleans (funds being all gone) and we came here to get fuel enough to carry us to Galveston. We arrived off the bar of this river too late on the night of 3rd October to come in, and toward morning we had a severe gale and sea from northeast, a little the worst many of us had ever seen. Now the Old *Zavala* stood it bravely, and after losing our rudder, best anchor and cable, the main mast, the guns and about 400 eighteen-pound balls and all our grape and canister overboard, cutting up the saloon, ward room, steerage and berth deck for fuel, we came in here all well and hearty, on the 7th of October, the hull of the vessel and the engines not being hurt at all.

(On this particular page in the original Tennison collection, the pencil had begun to fade, so Mr. Cornibe attempted to write over Tennison's handwriting, making it nearly impossible to read. It begins with an entry from Wednesday February 3, 1841, likely from a Galveston paper, announcing the return of the *Austin* from the Yucatan. The rest fades out quite badly. Then, a list of officers aboard the ship is presented, differing slightly from the last one of June, 1840. Some names are illegible and will be denoted as such.)

OFFICERS ON BOARD THE SLOOP-OF-WAR *AUSTIN*
BEARING THE BROAD PENNANT OF COMM. E. W. MOORE

1st Lieutenant - Downing H. Crisp
2nd Lieut. - Wm Seeger
3rd Lieut. - C. Leay
4th Lieut. - C. Cumming [hard to discern in original]
Acting Master - Wezman
Surgeon - J. B. Gardner
Purser - N. Hurd
Lieut. of Marines - T. W. Sweet
Commodore's Secretary - C. A. Cristman

Midshipmen:
James Mabry
C. Arcamble
J. H. Wheeler
Betts
Clement
W. W. McFarlane
Alf. Walke
Cornebay [A note from Dienst's source of the documents, E. J. Cornibe, appears here as "*My father Cornebay of Waco wrote this*," implying that it was Mr. Cornibe's father who wrote over Tennison's original pencil jottings as they faded during his possession of the documents.]

L. H. Smith
Joseph A. Fornish

(The following all appear in sequence in Tennison's documents and originate from issues of the *Galveston Morning Herald* bearing the dates shown.)

Jan. 9, 1841

The Texian schooner *San Bernard*, T. A. Taylor, commanding, came into port yesterday and anchored opposite the Navy Yard. The *San Bernard* is from a cruise along the Mexican coast and last from Tabasco, 12 days out.

Through the politeness of Lieut. Tennison, we are in possession of the following items of news. The steam ship *Zavala* and flagship *Austin* were at Tabasco when the *San Bernard* sailed. The former vessel has completely repaired the damages she sustained on the bar at the mouth of the Tabasco River, and would proceed to sea in a few days after the departure of the *San Bernard*, and intended towing the ship *Austin* over the bar.

The *Zavala* would proceed to Laguna for sufficient supply of fuel and then hold her course for Galveston.

The *Austin* would proceed on a cruise, touching at the Arcas Islands, and thence at this port. The schooner *San Jacinto* went ashore in a heavy gale (shortly before the sailing of the *San Bernard*) while anchored off the Arcas Islands, owing to the want of proper ground tackle and is a total wreck. We are happy to state, however, that no lives were lost. The officers and men attached to the navy at Tabasco were generally in good health. The Federalists still remained in possession of Tabasco.

Feb. 10, 1841

During the last cruise of the ship *Austin* she boarded a small schooner bound to Vera Cruz having on board the Federal General Lemus, prisoner to the Centralists. Commodore Moore demanded and obtained his release. The joy of Lemus and his wife was excessive. They were landed at Campeche and Lemus was in a responsible office under the new government of Yucatan.

March 20, 1841

The war schooner *San Bernard* returned to this port on the 18th, having left Vera Cruz on the 12th after delivering to the commander of the British sloop-of-war *Comus* the despatches brought out for the British Minister. The appearance of the *San Bernard* was by no means welcome at Vera Cruz, and the batteries were manned and forces marshalled to beat off the formidable enemy—a single schooner and 20 men. Eight boats with about 70 men each were prepared to attack her, but the interference of the British sloop prevented any difficulties. The courtesy of the British commander is spoken of in high terms by our officers. His name is E. Nepean and, as his vessel will be here in a short time, we hope our citizens will make some suitable return for his politeness. The delay of the *San Bernard* in delivering her despatches was caused by the loss of her foremast and the detention in making repairs at the Arcas Islands.

The *Zavala* was left at Laguna on the 1st inst., her supplies of fuel and provisions not then having arrived from New Orleans, though she is probably now on her way to this port.

April 10, 1841
From the Steam Ship of War *Zavala*

The steam ship *Zavala* arrived yesterday morning in five days from Yucatan. She has on board $8460 in specie, having received ten thousand dollars in payment of service rendered by our Navy in the taking of Tabasco, the balance being expended in the payment of debt contracted here.

At Yucatan everything was quiet. No standing army to make the civil authority subordinate to the military, as in many parts of Mexico; all kinds of religious worship was tolerated there.

Arista has joined Canales; but has no designs against Texas. He seems determined to overthrow the existing government.

We are assured by a passenger on board the *Zavala* that the Navy could, if permitted to make captures, not only defray its own expenses but support the government. How injudicious [is] the policy pursued by the President and Congress in relation to this arm of national defence.

Our informant tells us that Capt. Lothrop, who commands the *Zavala*, is one of the most indefatigable officers he has met with, and adorns by his qualifications, gallantry and judgement the station he occupies.

The *Zavala* arrived in port with not more than three barrels of coal remaining; an hour or two longer at sea and she would have been without steam.

She is in fine condition, particularly her engines, but her boilers are represented as being nearly burnt out and liable to explosion.

July 3, 1841

The war schooner *San Bernard* arrived here today with Judge Webb on board, and bringing intelligence that Mexico refused to receive or treat with him as an agent to procure the acknowledgment of the independence of Texas.

July 7, 1841

The war schooner *San Antonio* left this city on Sunday, for the purpose of surveying the coast, to commence at the Sabine Pass. List of officers on board the *San Antonio:*

[NOTE: The *Austin* was in ordinary at Galveston, like most of the remaining fleet, on orders from President Lamar, while Great Britain tried to negotiate an agreement between Texas and Mexico. So, we find the Commodore and some of his men aboard the *San Antonio* in the summer and fall of 1841 tending to coastal surveying.]

Commodore E. W. Moore
William Seegar, Esq., Lieutenant Commanding
A. A. Waite, 1st Lieutenant
C. Cummings, 2nd Lieutenant
R. M. Clarke, Surgeon
J. C. Wilber, Master
W. H. Sandusky, Draftsman
Monroe H. Dearborne, Midshipman
Charles S. Arcamble, Midshipman
William H. Allen, Midshipman
Edward Johns, Midshipman
John Thomson, Captain's Clerk
J. Brown, Boatswain

Passengers: Col. G. W. Hockley and servant

October 9, 1841

Lieut. Lewis left yesterday with despatches for Commodore Moore, who is down the coast surveying with the schooners *San Antonio* and *San Bernard*, the first commanded by himself and the latter by Lieut. Crisp. The despatches are supposed to request the return of these vessels for the purpose of [line illegible at fold of original MS].

The commissioners from Yucatan, employed to engage the cooperation of our navy with their government against Central Mexico, brought nine thousand dollars in specie with them, which was deposited in the Custom House in this city.

October 13, 1841

The schooner of war *San Antonio*, Commodore Moore, entered our harbor yesterday from the westward, having been engaged in surveying the coast in that direction. [NOTE: Per the *Matagorda Bulletin*, Moore was surveying in that vicinity prior to being recalled to Galveston.]

New appointments to the navy—W. T. Maury and J. F. Stephens, Pursers; Messrs S. L. Miller, A. Peyton and Thomas Henderson, Midshipmen; and Messrs Swisher, Archer and Roberts, Lieutenants.

The *Intelligencer* of Saturday has an article on the subject of the Navy, which says that the editor understands the arrangement made between our government and Yucatan to stipulate that "the State of Yucatan is to pay $8,000 per month thereafter," for the use and support of our navy, and "our government to furnish our best three vessels."

The same paper contained the following estimate which the editor says he believes to be very correct "and made by persons well capable of doing so."

Outfit and pay for one year of the ship, officers and men —$47,000

Same for steamer —$47,000

Repairs on *Zavala* to render her fit for service—$9,500

Outfit for one brig for one year and pay officers and men—$32,500

Same for two schooners—$32,500

Provisions for all of the above for one year—$45,000

Munitions of war—$7,000

TOTAL—$220,500

From which it would appear that Yucatan agrees to pay enough to keep a brig and two schooners in service. We do not know enough of vessels of war to state the precise quota of officers and men required by each, or whether there is an actual demand for the number usually allowed, but we discover a considerable discrepancy between the amount set down above and those laid before Congress by the Secretary of the Navy in 1839, as necessary for the support of the vessels pertaining to the navy. According to that gentleman, the pay of the officers, sailors and marines for the ship *Austin* is $45,586; for the steam ship *Zavala* $40,647; for a brig $36,767; for two schooners $45,558; and for provisions, munitions of war &c, exceed in the same proportion those published by the *Intelligencer*. We state this last fact because these last estimates were the basis of our remarks last week. We are inclined to think that many of the officers they [illegible at MS fold] with, and believe the estimate in the *Intelligencer* sufficiently high. We have been furnished with the following of all the principal officers now in the service and append the amount of pay according to the salaries established by law:

(1) Captain Commanding—$4,500
(1) Commander—$2,500
(13) Lieutenants, $1,500 each—$19,500
(4) Pursers, $1,500 each—$5,000
(2) Surgeons, $1,333.33 each—$2,666.66
(17) Midshipmen, $400 each—$6,800
(3) Lieuts. Marines, $652 each—$1,956

The law of last Congress declared that there shall be only eight lieutenants, ten midshipmen, one lieutenant and two pursers.

Saturday, October 22, 1841

Since the arrival of the Commodore, we have been informed that definite arrangements have been made for immediately fitting out the ship and the schooners for sea.

The *San Antonio* sailed yesterday with the Commodore and Col. Pereza on board, bound to New Orleans and thence to Campeche. We annex an estimate of above mentioned vessels for a period of three months, which has been kindly handed us from good authority:

Provisions for three months	$7,690
Clothing, same	$6,891
Rope, canvass, &c	$1,500
2,000 round shot, 24 lbs	$2,400
Freight on provisions, clothing	$400
Tobacco	$190
Enlisting 56 marines, $10 each	$560
Shipping 90 men, $20 each	$2,070
Passage of same at $9.50 each	$855
Pay of same for 3 months	$18,066.44
Total	$40,422.44

To defray the above expenses the government is to receive from the State of Yucatan the sum of $8,000 and in three months the sum of $24,000. From the charge of clothing, the sum of $5,500 will be returned.—$37,500

Leaving to be made up by this government the sum of $2,922.44.

It appears from the above that our government is to receive from the State of Yucatan $8,000 and "in three months the sum of $24,000." The [original MS ends here, as a page is missing.]

February 1, 1842

A letter has been received on board the ship *Austin* at Sisal dated January 18th, stating that Commodore Moore had gone to Merida, and while there the lieutenant in command of the ship became acquainted with the fact of the reunion of Yucatan with the Central Government. Apprehensive that Commodore Moore might be detained as a prisoner, he seized the Mexican commissioners as they were returning, and determined to detain them as hostages until the arrival of the Commodore. He, however, shortly after received a letter from him (Commodore Moore) stating that he would return to the ship in a day or two, and ordering the release of the commissioners, among whom, on the part of Santa Anna, were Generals Lemus and Anaya, hitherto supposed to be the most honest men among the Federalists.

Extract of a letter dated Sisal, 31st January, 1842, received by the schooner of war *San Antonio*:

> I have only time to write you a few lines by the *San Antonio*, which vessels leaves this morning for New Orleans via Galveston. The Texian fleet is here and has been cruising in the Gulf of Mexico and the Bay of Honduras since their departure from Galveston, without seeing the smallest sign of the enemy or anything belonging to them. By an arrival at this place yesterday evening from Vera Cruz via Campeche, we learn that the agents sent from the State of Yucatan to treat with Santa Anna for the reestablishment of the Constitution of 1824, together with the commissioners sent by him [illegible in MS] this government (Yucatan) were imprisoned on their arrival at Vera Cruz. They sailed from Sisal some three or four weeks since, for the object above mentioned, in the barque *Louisa*, which vessel the authorities at Vera

Cruz also attempted to detain. These same commissioners had been arrested before leaving Sisal by the First Lieutenant of the ship *Austin* (the Commodore being at that time at Merida) and kept on board the *Austin* until the Commodore was apprised of the fact and ordered their release. General Anaya who was with them was wise enough to keep aloof until he could learn upon what footing he would be received and thereby escaped the [illegible in MS] of his brother Mexicans.

From the same source we learn that the remains of the ill-fated Santa Fe expedition had arrived in the City of Mexico in chains, but through the remonstrance of foreign ministers, the chains were taken off the officers. What disposition is to be made of them is not known.

(taken from *Galveston Civilian*, February 22, 1842)

February 22, 1842

Lieut. Tennison of our navy arrived on Saturday in charge of the Mexican schooner *Progressa*, captured in sight of Vera Cruz on the 6th inst. by the sloop of war *Austin*... She is laden principally with flour and sugar. Lieutenant Tennison has despatches for our government. When the *Progressa* left, the schooner of war *San Bernard* was in chase of another Mexican vessel, which was...to have on board a large amount of specie belonging to the government. The *San Bernard* was to the windward of her and between her and the shore, and so certain was Commodore Moore of the prize that he would not think it worthwhile to join the chase with the sloop of war.

The *Progressa* had despatches for General Cos, who was at Tuspan, which were taken, and it was the intention of Commodore Moore to proceed to that place for the purpose of making a prisoner of the Mexican general.

A general officer was captured on the *Progressa*. When he saw the Texan flag run up, he tore off his epaulets, thrust them into his pockets. But it was no use. He was caught in the act. He will serve as security—in part—for some of the Santa Fe prisoners. If Cos is taken, which Lieutenant Tennison thinks is probable, he will be applied to the same use...Santa Anna has purchased an old English steam ship carrying 4 guns of an English system and if he has any spirit—with her and the New York brig—may offer Commodore Moore a fight. Nothing would be more welcome to the Tars.

(excerpted from *New Orleans Commercial Bulletin*, February 22, 1842)

February 22, 1842

THE MURDER AND THE MUTINY

We have received the following particulars of the mutiny on board the Texian schooner of war *San Antonio*. The men had by some means procured liquor and were nearly all of them in a state of intoxication, early on the evening of the 11th.

They were observed by the officers to be hatching a consultation on the forecastle, which circumstances induced them to prepare for an emergency, not suspecting however, that a mutiny was intended, but believing some of them might attempt to desert.

Soon after, Acting Master M. H. Dearborne, being then in charge of the deck, the Sergeant of Marines came aft with a file of men and said he wanted to go on shore. Dearborne replied that it was not in his power to give him permission, nor was it in the power of any officer on board to do so, and advised him to wait until the Captain came on board, when he did not doubt that everything would be satisfactorily arranged. The Sergeant was not satisfied but continued urging the point when Lieutenant Fuller, then in charge of the vessel, came on deck and enquired the cause of the disturbance, and at the same time seeing some of the men by the hatch ordered them aft to the gangway, and told them to state their complaints. They told him that they wished to go on shore and *would* go on shore. Lieutenant Fuller then ordered the Sergeant of the Marines to arm the Marine Guard, which was immediately done, and they were ranged on the starboard side of the quarter deck, the Sergeant arming himself with a Colt's pistol and a tomahawk, and in all probability then furnished the crew with arms.

The Sergeant then came up to Lieutenant Fuller under the pretext of reporting the guard under arms, raised the tomahawk and struck without hitting him. Lieutenant Fuller then drew his pistol and in the act of raising it was shot by the Sergeant and killed instantly, the ball passing through his right arm near the shoulder and entering his lungs. He fell and while on the deck was beaten with muskets and cutlasses, the other officers standing around him endeavoring to ward off the blows. While engaged in this, Midshipmen Allen and Odell were wounded, the former by a ball through his foot and the latter by a ball through his thigh.

The mutineers after committing this murderous act, seemed satisfied, and told the survivors to go aft; if they did not, they also would be killed, some of the mutineers, at the same time seizing them and pushing them aft and down into the cabin, securing the doors. They then, it appears, armed themselves, lowered the boats, and nine of them crossed the river in one of them, and the remainder landed at Algiers in the other.

Captain Day of the revenue cutter *Jackson*, having heard the noise and the report of pistols, manned and armed his boats and pulled toward the *San Antonio*. The officers in the cabin hearing the noise of oars came on deck and hailed him, requesting his assistance, which was promptly rendered. Much credit is due to this gentleman and to his officers and men.

His Honor, the Mayor, with his usual energy, assisted by an efficient police, has succeeded in arresting six of the mutineers, and there is no doubt but the remainder will soon be brought to justice. The following is a list of those taken: Seymour Oswald, Sergeant of the Marines; William Simpson, Corporal of Marines; Thomas D. Shepherd, sailmaker's mate; Ed Keenan, Benjamin Pompilly and Edward Williams.

The remains of the late Lieutenant Fuller were interred yesterday afternoon, in the protestant cemetery, by his brother officers, Captain Day and the officers of the cutter *Jackson* and his particular friends.

(Galveston newspaper, excerpted in April 13, 1842 *Telegraph*.)

The man of war schooner *San Bernard* arrived on Sunday from the coast of Mexico. We have seen a letter from an officer on board the sloop of war *Austin*, which says that Commodore Moore will spare no efforts to keep the Mexican navy off our coast, and that although the enemy may destroy his ships, they shall never take her as long as she floats.

On Monday Lieutenant Lansing came in with another prize—a schooner called *Dos Amigos*, captured by the ship *Austin* off Tuspan. She is freighted with salt, and a fine little vessel said to have been built in the United States. This is the third capture made by our navy in little more than a month, and we understand that the Mexican vessels are now nearly as scarce as sperm whales within the Gulf.

(Galveston newspaper, reprinted in July 27, 1842 *Telegraph*.)

The Senate has confirmed the following appointments in the Navy. It will be remembered that although nearly all of these officers have been a long time in service and some even retired, the Senate has never voted upon their nominations until now.

E. W. Moore, Post Captain Commanding
J. T. K. Lothrop, Commander
D. H. Crisp, Lieutenant
W. C. Brashears, Lieut.
William Seeger, Lieut.
A. G. Gray, Lieut.
A. J. Lewis, Lieut.
J. P. Lansing, Lieut.
A. A. Waite, Lieut.
George C. Bunner, Lieut.
William A. Tennison, Lieut.
William Oliver, Lieut.
Cyrus Cummings, Lieut.
C. B. Snow, Lieut.
D. C. Wilber, Lieut.
M. H. Dearborn, Lieut.
Thomas P. Anderson, Surgeon
R. M. Clarke, Surgeon
J. B. Gardiner, Surgeon
Norman Hurd, Purser
F. T. Wells, Purser
J. F. Stephens, Purser
W. T. Brannum, Purser

(Galveston newspaper, August 10, 1842.)

The war schooner *San Bernard* arrived here on Saturday.

(Galveston newspaper, September 14, 1842.)

The officers, seamen and marines on board the sloop of war *Austin*, at New Orleans, held an election on the 5th inst. for Senator and Representative for this county and district, which they claimed the right to do, under the law authorizing citizens of a county, when absent in the public service, to open polls under the superintendence of three of their members, and forward the results to the Chief Justice of this county. Forty-two votes were cast for Representative—41 for Major Bache and one for Colonel Potter—and 40 for the Senate—all for General Baker.

Captain R. Oliver of the Texas Marine Corps died at New Orleans on the 11th of October.

Midshipman Culp of our Navy has been killed in a duel.

(Galveston newspaper, April, 1843)

Commodore Moore sailed from the Mississippi Pass Wednesday evening, the 19th instant, with the sloop of war *Austin* and the brig *Wharton*. Col. W. G. Cook is on board. The *Austin* has one hundred and forty-six men and the *Wharton* eighty-six, all told. The ship carries twenty guns, medium 24s, and the brig sixteen 18-pounders. They are both provisioned with a new species of shell shot, which are pronounced far superior to the Paixhans. It was expected, when the vessels sailed, that they would touch here, but it is now believed that they have gone directly to Yucatan. We shall await with interest accounts of their operations there.

[NOTE: Here begins the final cruise of the Second Texian Navy, with Colonel Morgan on board, for which Moore will be condemned by Sam Houston.]

Letter from Commodore Moore to the mayor of New Orleans, August 12, 1842, requesting the release of the *San Antonio* mutineers to his custody so that they may be brought to justice. Dienst Collection, Briscoe Center for American History.

(*New Orleans Commercial Bulletin*, April 29, 1843)

Below we publish a letter from Commodore Moore written before the departure of the Texian squadron from the Balize given the insult of the late courts martial held on the ship *Austin*, the details of which were published in the city paper. It will be seen that four of the mutineers have been sentenced to death, three to be flogged and one discharged. The Commodore stated in the following letter that he will carry out the sentences of the courts martial in these cases in a few days. We are in possession of information which induced us to believe that the terrible penalty of the law they violated has already been paid by these miserable men, Landois, Hudgins, Allen and Simpson undoubtedly expiated their offences yesterday noon, at the yard arm of the ship *Austin*. These men were patiently tried, and no doubt of their participation in the murder of Lieutenant Fuller is entertained by those who heard the evidence presented to the court. The fearful example of their execution, so justly ordered, will exercise a most beneficial effect in preserving the discipline of the navy of Texas.

Texas Sloop of War *Austin*
off N. E. Pass of Mississippi River
April 18, 1843

To the Editors of the *Tropic:*
Herewith I forward you the sentences of the Courts Martial convened on board this ship at New Orleans.
No. 1. Frederick Shepherd—*not guilty*
No. 2. Antonio Landois—guilty of all the charges and specifications and sentenced *to suffer death* unanimously
No. 3. James Hudgins—guilty of all the charges and specifications and sentenced *to suffer death* unanimously
No. 4—Wm. Barrington—*sentenced to receive one hundred* lashes on the bare back.

No. 5. Isaac Allen—guilty of the 1st and 3rd charges, sentenced *to suffer death* unanimously.

No. 6. John W. Williams—guilty of the 3rd charge and sentenced to receive fifty lashes but recommended to mercy.

No. 7. Edward Keenan—guilty of the 3rd charge and *sentenced to receive one hundred* lashes.

No. 8. William Simpson—guilty and sentenced to suffer death—three members for the sentence of death, *one* member for the sentence of death and recommendation to mercy, and one member for the sentence to receive one hundred lashes with the cats.

The following was the rating of the respective men on board the *San Antonio*:

No. 1. Frederick Shepherd—Seaman
No. 2. Antonio Landois—Marine
No. 3. James Hudgins—Seaman
No. 4. Wm. Barrington—Seaman
No. 5. Isaac Allen—Seaman
No. 6. John W. Williams—Seaman
No. 7. Edward Keenan—Cook
No. 8. William Simpson—Capt. Marines

I have numbered the men in the order that they were tried.

By the evidence presented to the Court, a regularly concerted plan had been forming for some time, to seize this ship and the schooners *San Antonio* and *San Bernard* (the only vessels at sea) and run them into Vera Cruz. I have thought it best to give you this information, as in the course of human events we might all go to the bottom.

I enclose you a copy of the charges against Frederick Shepherd and Antonio Landois, all the rest are the same as those against Landois.

The sentence of the Court in the case of Midshipman R. H. Clements, will require the action of the President of Texas—the others I will carry out in a few days myself.

We sail first for Galveston, where I contemplate stopping for a few hours, when I will sail direct to attack the squadron off the coast of Yucatan.

Yours truly,

E. W. MOORE

(likely the New Orleans *Tropic* article referenced in above article. Tennison erroneously dates this entry "May 10th, 1840" but since it had to precede the above article, it was likely published the week of April 23, 1843.)

MUTINY ON BOARD THE SAN ANTONIO

The court martial was convened on board the Texian sloop of war *Austin*, for the trial of Fred Shepherd and eight others, on charge of mutiny on board the Texian schooner of war *San Antonio* on the 11th of February 1842, whilst lying in the Mississippi River, opposite this city.

In the mutiny, Lieutenant Fuller was killed and two midshipmen were wounded, the particulars of which were published at the time. Most of the mutineers have been in the city prisons from that period until some time last week when they were delivered to the Texian Commodore upon the requisition of President Houston. The Court assembled at 10 o'clock, composed of the following—Commander J. T. K. Lothrop, President; Lieutenants A. G. Gray, J. P. Lansin, Cyrus Cummings and Wilber, members; Surgeon T. P. Anderson, Judge Advocate.

The Court proceeded to the trial of F. Shepherd, boatswain of the *San Antonio*, and after the examination of several witnesses, J. D. Shepherd, one of the mutineers, was called, who turned State's evidence, upon a grant of a free pardon by the President. But for this and similar testimony, the prosecution might have failed, as the principal witnesses have perished upon the ill-fated *San Antonio*, which was lost in the Gulf on the 5th of September 1842.

Joseph D. Shepherd was sworn and testified to the following effect: That the mutiny had been projected while the *San Antonio* was lying at the island of Mujeres, off the eastern coast of Yucatan, in January 1842. He stated that the crews of the *San Bernard* and *San Antonio*, while on shore for water, drew up articles of agreement, for the purpose of seizing one of the vessels, the *San Antonio*, with the view to take her to Vera Cruz and sell her to the Mexican government. The other vessel was to be left with her officers on board if they yielded quietly, but in case they resisted, they were to have been put to death.

Before the mutineers could perfect their schemes, they were prevented from going ashore and they then proposed deferring their arrangements until they reached Sisal. On arriving at Sisal, they found the sloop of war *Austin* at that place and were consequently again obliged to postpone the accomplishment of their designs until another time. The *San Antonio* was ordered to New Orleans, but owing to the number of passengers on board, they were unable to execute their designs during the voyage.

On the night of the 11th of February, 1842, after the vessel arrived at New Orleans, about half-past 8 o'clock, the men had secretly obtained liquor and agreed to take the boats and go ashore. They had got possession of some of the small arms, and demanded from the officer of the deck, Mr. Dearborn, master, permission to leave, which was refused. Lieutenant Fuller came on deck and ordered the Sergeant of the Marines to arm the Marine Guard, which was done. Several pistols were discharged, and Lieutenant Fuller was killed, receiving a pistol ball in his shoulder which lodged in his lungs, causing immediate death. Midshipman Wm. H. Allen was shot in the foot, and was knocked down by

Antonio Landois, with a musket. Landois attempted to run the bayonet through him but was prevented by the witness, J. D. Shepherd. Midshipman T. H. Odell was also wounded in the leg with a pistol ball.

The men then took the boats and left the vessel, except the [illegible in MS], Frederick S. Shepherd, who remained on board. The following men were arrested next day, viz: Seymour Oswald, Sergeant of Marines; Benjamin Pompilly, Joseph D. Shepherd, Wm. Simpson, Edward Keenan, James Hudgins, Wm. Barrington, Isaac Allen and J. W. Williams. Seymour Oswald escaped from the parish prison. Benjamin Pompilly died in prison and confessed on his death bed that he had killed Lieutenant Fuller. Antonio Landois was arrested in August following, in a swamp near New Orleans.

Here the testimony closed, and the Court adjourned to this morning at half-past 9 o'clock.

(*New Orleans Tropic*, May, 1843, specific date unknown. Also published in the *New Orleans Civilian* on May 27, 1843. Although signed simply "J" in Tennison's transcription and in newsprint alike, it was likely penned by J. T. K. Lothrop, commander of the *Wharton*.)

Texian Brig of War *Wharton*
Off Campeche, May 10th, 1843

By the last arrivals from this quarter at your port, I presume you were placed in possession of full information relative to our movements up to the 4th or 5th instant.

Since that period, nothing of particular importance has occurred. We have tried to force the Mexicans into a fight but they will not "come to the scratch." On the morning of the 2nd instant at 4 o'clock, a fair breeze springing up, we got underway and stood out for the Mexican squadron. The land breeze, by which alone we could hope to reach them, died away and we were compelled to beat in to our anchorage. The Next morning at five o'clock the breeze again set in, when we once more made sail for the enemy.

We came near enough to their ships to exchange a few shots and we are in hopes, from the confusion observable on one of them, that one of our balls did considerable damage. They soon played the old game and retreated. The object of the Mexicans, as we plainly understand, is to induce us to leave our anchorage with the morning breeze—and as the calm usually comes on a little before mid-day, they hope to catch us powerless, and to use us up by the aid of steam, in the most summary manner. To this operation, as you may well imagine, we all object—Give us a good wind and our word for it, the Texian Navy will prove itself true to the core—we await only the favors of fortune, due to those who diligently seek them.

On shore the operations are of an odd nature. As near as we can judge, the Mexicans in possession of the heights were reinforced today by about two thousand men, and beyond all doubt a fierce attack upon the city is contemplated. The Campechanos expect reinforcements from Merida every hour, and whether their expectations are fulfilled or not, they will unquestionably hold out against the enemy.

Midshipman Faysoux, who you have learned was slightly wounded in the actions of the 30th ult., has perfectly recovered and is now on duty. All our officers are in the finest health and spirits, and anxiously desire "a fair fight and no favor." We pray for an early chance at the enemy. You shall hear from me at every opportunity.

Very truly yours,

J.

(Likely from the *Picayune*. Most Texas papers carried Commodore Moore's minutes of the engagement in their early June issues, and a few carried this letter as published in the *Picayune*. The *New Orleans Republican* carried it on May 27.)

Extract from a letter written by an officer on the brig *Wharton* dated May 17th 1843

On the morning of the 16th (says the writer), Commodore Moore made signal to attack the enemy, then about a mile and a half to windward, or rather seaward, of us. This we soon answered and brought our long Tom to bear on the iron steamer *Guadaloupe*, the Commodore at the same time engaging the *Montezuma*. The action here continued hot and heavy for about two hours, changing our position but very little, and the steamers occasionally hauling off to repair damages, or probably to try and get at such a distance as would be too great for our shot; but as they shortly afterwards returned to the fight, I am induced to believe they hauled off to repair damages.

The sea breeze about this time sprang up and the action again became general—the Commodore engaging the *Montezuma* and our little beauty the *Guadaloupe*. The action now, as well as the day, became very hot; shot flew in every direction but none of them took the least effect; while our shot, I am sure, must have done immense damage. We have not been able to ascertain to a certainty, but all are of the opinion on board that we damaged the machinery of the *Guadaloupe* very materially. One thing is certain, she was enveloped in her own steam for a considerable time, and was seen afterwards to work with one wheel.

About half-past 1 p.m., the attention of the steamers seemed directed to the Commodore and they rarely deigned to give us a shot, although we kept up a very brisk fire on both. About 2 o'clock, one of our seamen was blown to atoms by the gun at which he was engaged. He was ramming home the cartridge and the captain of the gun not seeing him at the muzzle, fired, and blew the man to pieces. From this time till 3 o'clock, the action became general and very hot. The Commodore received several shots in his hull and bulwarks.

At 3 the steamers hauled off, and the Commodore, having received a shot between wind and water, which caused his vessel to leak much, wore ship and stood back for the anchorage. He made signal for us also to withdraw from the action. We gave the *Guadaloupe* a parting broadside and followed in the wake of the ship. At 4, came to anchor off Campeche, where we learned that the ship had received fifteen shot. She was shot in almost every direction, below and aloft; yet what is the most remarkable thing is she did not lose a single spar. She had two killed and twenty five wounded; among the latter is Lieutenant Wilber and Midshipman Bryant. A number of amputations have taken place, a great pity, as a better and braver crew never trod a deck.

It was with great reluctance we withdrew from the action, but the magazine of the flagship had 25 inches of water in it, and as they had no powder it was useless to remain a target for the enemy. The Mexicans seemed very glad that the day's work had terminated, as they did not make the least effort to follow up the engagement. The loss on their side must be very great, as at the commencement of the action, their decks were lined with men and I am quite confident that if the action lasted one hour longer, and we could get a little closer to them than we were, we would have captured them.

(Most Texas and certainly the New Orleans papers carried Moore's description of the Battle of Campeche, shown below. The *Telegraph* printed it in their June 7, 1843 number, sans the official minutes and Surgeon's report. The *Galveston Civilian* carried it in full in its June 10, 1843 edition.)

Texas Sloop of War *Austin*
Off Campeche, May 19th, 1843

Messrs Editors—I wrote you on the 5th inst., enclosing you a copy of the minutes of the action of the 30th April.

I herewith enclose you a copy of the minutes of our action of the 16th inst.—the wind was so light, that at no time were we nearer than a *mile and three quarters*—The advantage of steam and heavy guns is tremendous, particularly in the Gulf of Mexico, in the summer, where there is so much light weather. The whole fire of both Steamers was directed at this Ship; not a shot struck the Brig. We fired over *five hundred* cartridges, and the long gun from the *Wharton*, (obtained since we arrived here,) was fired *sixty-five times*, and repeatedly with great effect. When the sea breeze came in, the Brig was caught aback, and before she got round on the same tack, we were some distance ahead, because we could not heave to, as we would have run the risk of being raked; she however made sail, and soon got within gun shot again. I forgot to mention that I have obtained, since my arrival, *two* long 18-pounders, which have been of great service.

Our crew behaved nobly! A finer set of men were never on board ship; and as to the officers under my command, it would be impossible for me to express to you my admiration of their conduct and bearing.

The *Guadaloupe* was very much crippled, and when she wore to, to stand in for anchorage, she did not move her wheels for over *forty* minutes; she was however to windward and we could not get up to her; *one* of the shell shot struck the *Guadaloupe* abaft the wheel, which must have done her considerable injury. I have fired but very few of them, owing to the distance that they have always kept from us; I am reserving them for close quarters.

On the 17th and 18th there was brisk firing kept up on both sides on shore, and the gun boats took a hand—the Mexicans having taken a position in the suburb of San Roman, which they were compelled to evacuate this morning. I was on the walls yesterday morning for about an hour, and the musket balls were whizzing a small few, and many were injured on the other side. I imagine they were not less than *five or six hundred* yards apart.

I also enclose you a copy of the Surgeon's report of killed and wounded. Frederick Shepherd was one of the crew of the *San Antonio*; he was confined on board this ship from the 11th March, 1842, until his acquittal and release; he was captain of No. 5 gun, and most nobly had the poor fellow redeemed his character from the charge of participating in the mutiny on board that ill-fated vessel—for a better I had not on board the vessel.

The wounded men are doing well; several of them are in the hospital at Campeche, and the Governor is very urgent that they should all be sent on shore; (two men have had their arms amputated, and one a leg—Thomas Barnet and John Norris an arm, and Owen Timothy a leg;) but the Surgeon prefers keeping them on board, at least for the present.

Those 68 pound balls are tremendous missiles, and the way they did whistle or rather hum over our heads was a caution, I tell you. They fired a great many over the poop where I was standing, and several of them were disposed to be *rather too intimate.*

I will be ready to give them another chance in a few days, but I will wait for a strong breeze, and if I *can get* near enough to use our shell shot, I feel confident that they will tell a big tale in a few minutes.

A fisherman came in this afternoon and said that he had been alongside of the *Guadaloupe* today and that she had *forty-seven* men killed and *thirty* wounded; he also said that nearly all the men were on board the *Guadaloupe* for the purpose of boarding this ship, which, by the way, they had a fair chance of doing, for, during the *four hours* fight, we chased them *not less* then *twelve or fourteen* miles.

Yours truly,
E. W. MOORE

May 20 at 3 o'clock p.m.—Nothing done in the way of fighting on either side since yesterday. I forgot to mention that we have the *Guadaloupe's* flag staff on board this ship.

E. W. M.

Texas Sloop of War *Austin*
Off Campeche, May 16th, 1843

At 4:30 a.m. called all hands and piped the hammocks up—at 4:45 called all hands "up anchor"—at 5, set the top-sails—at 5:20, made signal No. 20 to Yucatan Squadron—at 5:25, made signal No. 406—at 5:30 weighed anchor and filled away, head S. and W., wind S.E. and light, *Wharton* in company; Yucatan Squadron getting underway—at 6, set topgallant sails—at 6:12 made signal No. 10 to Yucatan Squadron—at 6:20, made signal No. 77, beat to quarters and cleared ship for action; enemy's squadron underway, bearing W.S.W., distant 5 miles—at 6:45, enemy stand-ing off, beat the retreat and piped to breakfast—at 8 hoisted Texas ensign at the peak, and broad pennant at the main—at 10 nearly calm—at 10:40, enemy's squadron hoisted their colors, *Guadaloupe* hoisting at the same time English ensigns at the fore; the *Montezuma*, English at the main and Spanish at the fore, and stood towards us; beat to quarters, hoisted English and American ensigns

308

at the fore and Texas at the mizen—at 10:55, ship's head S.W. on the starboard tack, *Wharton* about one fourth of a mile astern, Yucatan squadron close in shore, enemy about two and a half miles off our larboard bow, commenced firing at us, most of their shots falling short—at 11:05 made signal No. 96 and fired larboard broadside; the medium twenty-fours not reaching, ceased firing with all except the long eighteen, *Wharton* commenced firing at the same time—at 11:18 second shot from the long gun cut away the *Guadaloupe's* flag staff which fell overboard with the ensign, when the crews of both vessels gave three hearty cheers; *Guadaloupe* hoisting another ensign at the main gaff—at 11:37 a thirty-two pound shot from schooner *Eagle* came through the hammock netting larboard side, over No. 6 gun, struck the combings of the steerage hatch, rebounded, struck the deck and passed out No. 7 port, wounding three men—closing upon enemy, commenced firing the medium guns.

At 11:40, a shot from the *Guadaloupe* cut away the starboard main topgallant breast backstay, after shroud main topgallant rigging, starboard main royal lift and halliards, and passed through the main topgallant sail—At 11:43, a shot from the *Guadaloupe* cut the starboard fore topgallant studding sail yard in two, the sail being in the top—at 12:20 the sea breeze setting in, but very light, the *Montezuma* being on our larboard quarter, set the foresail, put the helm up, squared the yard, manned both batteries and ran directly between both vessels, trying to bring them to close quarters, giving them our broadsides as the guns bore; upon which the schooner *Eagle* tacked, made all sail and stood off to the southward, and did not come into the action again; the steamers finding that we were bringing them to close quarters, and the wind being light, paddled on and took their position on our starboard bow. At 12:50, a sixty-eight pound shot from the *Guadaloupe* came through hammock settings, starboard side, over No. 7, passed out larboard side, carrying away forward port stanchion No. 9 port and the mizen channels forward of the port, with two chain plates—Up to 1:42, the firing continued on both sides, enemy's shots passing between our masts and over the poop—At 1:42, a sixty-eight pound shot from the *Guadaloupe* cut away the starboard main brace, bumpkin, &c. At 1:45, a shot from the *Guadaloupe* cut away the fourth shroud of the starboard main rigging, starboard main truss and foot rope of the main topsail—At 2, a shell exploded overhead, cutting the main royal mast badly, and several ropes—At 2:10, a forty-two pound shot from the *Montezuma* struck just forward of No. 9 gun, starboard side, passed through the waterways and deck, into the ward room through No. 3 state room, Purser's store room, and lodged in the armory, wounding two men at No. 9; Thomas Norris, one of the wounded, returned to his quarters as soon as his wounds were dressed, and in a few minutes his left arm was shot off.

At 2:24 a shot from the *Guadaloupe* passed through the ensign at the peak—at 2:25 a sixty-eight pound shot from the *Guadaloupe* struck the edge of the copper under No. 1 gun, broke the planking and rebounded, causing a bad leak which was immediately plugged—at 2:26 a shot from the *Guadaloupe* cut away the third shroud of the starboard fore rigging and starboard futock shroud—at 2:24 a sixty-eight pound shot from the *Guadaloupe* passed through the hammock nettings over No. 7 gun, killing one man and wounding Lieutenant Wilber, Midshipman Bryant, S., Hubbard, Captain's clerk, and four men—at 2:35 a sixty-eight pound shot from the *Guadaloupe* struck sill of No. 5 port, starboard side, passed through and carried away both alxetrees of the gun carriage, ripped up the deck three feet, injured the main topsail sheet bits and main mast fife rail and stopped on deck, killing Capt. No. 5 gun and wounding four men—at 2:36 a shot came through the hammock nettings over No. 6 gun starboard side, killing one man and wounding four—at 2:40 a shot cut away starboard main topgallant-backstay—at 2:42 a shot cut away 2nd shroud starboard mizen rigging, mizen topgallant halliards and larboard main brace—2:45 a sixty-eight pound shot came through the bulwarks starboard side abaft No. 9 gun, above the pin rail, wounding 2 men and passed out opposite port—at 3 p.m. the breeze freshening and all the weather main topgallant rigging being cut away, one gun of starboard battery disabled, wore ship to engage the enemy with the larboard battery, *Guadaloupe* ceased firing and still standing on the starboard tack, being too leeward and not being able to bring the enemy to close quarters, made signal No. 81 and kept off for Campeche, *Wharton* in company, Yucatan Squadron out of gun shot dead to leeward.

JAMES W. MOORE, Secretary

RETURN OF KILLED AND WOUNDED
On the Texian Sloop of War *Austin*, May 16, 1843

KILLED

Frederick Shepherd, Boatswain's Mate
George Barton, Landsman
William West, Ordinary Seaman

Wounded Dangerously

John Norris, Seaman
Dick Stretchout, Boy
Thomas Barnet, Boy
George Davis, Marine
Owen Timothy, Landsman
Asa Wheeler, Marine

Wounded Severely

D. C. Wilber, Fourth Lieutenant
A. J. Bryan, Midshipman
Thomas Atkins, Seaman
George Firur, Landsman

Wounded Slightly

John Noland, Seaman
Wm Cole, Carpenter's Mate
Daniel White, Carpenter's Mate
Sildon Hubbard, Captain's Clerk
Joe Murphy, Landsman
Charles Hanson, Landsman
Nicholas Brady, Steerage Boy
John Little, Seaman
John Duston, Seaman
George Hamilton, Seaman
Wm. Barrington, Seaman
J. P. Landis, Purser's Steward

Total Killed:	3
Wounded Dangerously:	6
Wounded Severely:	4
Wounded Slightly:	12
Total:	25

THOS. P. ANDERSON, Surgeon

May 20th, 1843.—Several of the wounded have already returned to duty, and, with the exception of six, they will all be on duty in a few days.

E. W. MOORE

(Originally published in the *New Orleans Tropic*, likely by way of a Galveston paper. The *Telegraph* printed it in their June 7, 1843 number along with Moore's description of the Battle of Campeche, previous.)

The following is an extract from a letter written from Col. Morgan to a gentleman now in Galveston.

Campeche, May 20th, 1843

Commodore Moore fought his ship well, and but for a shot striking the *Austin* between wind and water, would have captured the famous steamer *Guadaloupe* beyond a doubt! The damage done the latter is almost irreparable: although she is of iron, she is nearly torn to pieces! This *Guadaloupe* is the flagship and now commanded by Don Thomas Marine, who sent Commodore Moore a challenge to come out and fight him in "three fathoms water." The challenge was in print, and in rather vulgar, braggadocia style, dated on the 15th but not received in Campeche until late in the day of the 16th. Commodore Moore never saw it until after the action, nor heard of it, but Don Thomas must have thought him very prompt in accepting it, for at *daylight* on the 16th, he was underway, going out to fight the fleet! So you will see the Commodore not only gave Don Thomas battle in *three* fathoms water, but chased him off into 20 fathoms and to windward out of reach!

Don Thomas wanted to do something handsome, and to make the capture of the *Austin* certain, he took on board 300 additional men, for the purpose of boarding; but finding it rather warm work at long shot, he gave up the contest...

[Col. Morgan's letter continues in the newspapers, but Tennison left it off here and picked up the following, which he dates July 19, 1843, and was most likely an editorial appearing in a Galveston paper. Six months prior, Congress had passed a secret act to authorize the sale of

313

the vessels of the Republic. The response of Galveston was overwhelmingly one of disapproval, so much so that when it came time for the ships to be sold to the highest bidder in November 1843, Galvestonians prevented bids from being placed, leaving Texas in possession of a fleet which saw its last action at the Battle of Campeche.]

THE NAVY—Many rumors have been circulated relative to the secret law for the future disposition of the Navy, and although the precise text of this enactment has never been made public, the prevailing impression apparent to us is that the sale of the vessels has been ordered, without regard to our attitude towards Mexico. If this is the complexion of the law, we cannot regard it as politic. The navy still possesses four good vessels. The two just returned from active service have demonstrated their efficiency in action, and came back in the best possible condition—not a whit the worse for their rough encounters with the enemy. The two which remain in port are undergoing repairs and may be speedily placed in a similar condition.

As long as they remain at the disposal of the government, there is no danger of an attempt on the part of Mexico to blockade our ports, in case negotiations for peace should be broken off, of which these would be at risk; in case any piece of unexpected good fortune should place her in possession of means sufficient for the prosecution of the war on a respectable footing. Americans, as all, are for peace, and sanguine of its attainment, we are yet satisfied that the best means of hastening its consummation is by continuing to impress the enemy with our own strength and ability to carry on the war, both defensive and offensive. To dispense with our Navy while she maintains hers would be to give her a decided advantage, as well as to remove the fear of molestations and the necessity of maintaining a force for defense in her coast towns.

It is but that the government is too poor to keep all our vessels in commission. But the expense of keeping them in repair and ready to be put in service in an emergency would not be great. Their complements of men could soon be made up of volunteers.

There are a number of...brave and war-tried officers still in the Navy, who have served the country faithfully for years, and received scarcely any reward except their... honor, who may yet be induced to remain in the service until they see the independence of the country established upon a firm basis. Many of them, if now discharged, may find themselves without means of employment, and left to struggle with adversity and want; the recollection of their unrequited services and sufferings will leave them with no favorable recollections of the gratitude of the nation or disposition to draw their swords again in its defence, should war assail. On the establishment of peace, after the American revolution, the only persons who appeared to suffer were those who had [illegible in MS] money or carried arms in defence of the country. We hope that such will not be the case here. Sufficient crews to keep the vessels in repair in ordinary could, no doubt, be retained out of the men already in service.

We regard this as an important matter. The present is a crisis in our affairs which has not happened before and may not again. If skillfully used, we have no doubt of a full guarantee of peace within a few months. The expenses of supporting the Navy for that time, at the highest rate, could be well afforded—But it may be made light. Sailors are familiar with rough fare, and we believe the farmers of the country would willingly contribute enough corn and beef, of which there is a surplus in the country, to support them.

Under the heading of "June 1846" is Tennison's final entry, chronologically, which I believe is the only other original entry in the collection other than the very first entry. It reads thus:

The Texas Navy was transferred by Lieut. Com'd. W. A. Tennison, agent on the part of Texas to [illegible in MS], agent on the part of the United States in June 1846.

Tennison remained in command until September when he found that he was not recognized as an officer by the U. S. government. He gave up charge of the Navy to Midshipman C. J. Faysoux and left the State of Texas without commission or money, and beg[ged] his passage as far as New Orleans and then to Washington City—and that after serving Texas over ten years constant from February 1836 to September 1846.

Sam Houston vs.
Edwin Ward Moore

As discussed in "Thumbnail Sketch of the Second Texian Navy," Sam Houston and Edwin Moore were not chums. Moore's moxie irritated Houston, and the charges levelled by Houston followed Moore around like a black cloud. When Moore attempted to fight for a commission in the United States Navy, the charges resurfaced. Over and over again, he was forced to say, "I was acquitted! What more do you want from me, people?" Repeatedly, he'd bring out the evidence and relive his court martial. He published, in September of 1843, a 200 page pamphlet, *To the People of Texas*, in which he carefully presents his case to the people and to the Congress, that he was illegally dismissed by Houston and deserved a court martial. Congress acted and made that tribunal happen, even though it could not be presided over by naval personnel. He would be acquitted of all but minor charges and the Congress sustained him in his naval commission. Sam Houston flatly disagreed with the verdict and with Congress.

What follow are various pamphlets produced by Moore, trying to fend off the charges that continued to haunt him, as well as his half of the pamphlet war that ensued when he and other Texian officers attempted to secure commissions in the United States Navy. The officers of that body, in trying to guard their own commissions, argued that the word "navy" as used in the annexation treaty referred only to the vessels, and not the men who peopled them. They also argued that Moore did not qualify even if the "navy" included the officers, since he was kicked out of his job by President Houston. Moore passionately and at length defended his reputation. He spent much of the rest of his life fighting these exhaustive battles.

PROCLAMATION.
By the President of the Republic of Texas.

Whereas, E. W. Moore, a post captain commanding in the navy of Texas, was, on the 29th of October, 1842, by the acting Secretary of War and Marine, under directions of the President, ordered to leave the port of New Orleans in the United States, and sail with all the vessels under his command to the port of Galveston in Texas; and whereas, the said orders were reiterated on the 5th and 16 November, 1842; and whereas, he, the said post captain, E. W. Moore, was ordered again, on the 29th December, 1842 to "proceed immediately and report to the department in person;" and whereas, he was again, on the 2nd of January, 1843, ordered to act in conformity with previous orders, and, if practicable, report at Galveston; and whereas, he was again on the 22nd of the same month peremptorily ordered to report in person to the department and to "leave the ship *Austin* and brig *Wharton* under the command of the senior officer present;" and whereas, also commissioners were appointed and duly commissioned, under a secret act of Congress of the Republic, in relation to the future disposition of the navy of Texas, who proceeded to New Orleans in discharge of the duties assigned them; and whereas, the said post captain E. W. Moore has disobeyed and continues to disobey, all orders of this government, and has refused and continues to refuse, to deliver over said vessels to the said commissioners in accordance with law; but, on the contrary, declares a disregard of the orders of this government, and avows his intention to proceed to sea under the flag of Texas, and in direct violation of said orders, and cruise upon the high seas with armed vessels, contrary to the law of this republic and of nations; and whereas, the president of this Republic is determined to enforce the law and exonerate the nation from the imputation and sanction of such infamous conduct; and with a view to exercise the offices of friendship and good neighborhood towards those nations whose recognition has been obtained, for the purpose of according due respect to the safety of commerce and the maintenance of those most essential rules of subordination which have not heretofore been so flagrantly violated by the subaltern officers of any organized government, known to the present age, it has become necessary and proper to make public these various acts of disobedience, contumacy and mutiny, on the part of the said post captain, E. W. Moore; therefore:

I, Samuel Houston, President and Commander-In-Chief of the army and navy of the Republic of Texas, do, by these presents, declare and proclaim, that he, the aforesaid post captain, E. W. Moore, is suspended from all command in the navy of the Republic, and that all orders, sealed or otherwise, which were issued to said post captain, E. W. Moore, previous to the 29th October, 1842, are hereby revoked, and declared null and void, and he is hereby commanded to obey his subsequent orders, and report forthwith, in person, to the head of the Department of War and Marine, of this government.

And I do further declare and proclaim on failure of obedience to this command, or on his having gone to sea, contrary to orders, that this government will no longer hold itself responsible for his acts upon the high seas: but in such case, requests all the governments, in treaty, or on terms of amity with this government, and all naval officers on the high seas or in ports foreign to this country, to seize the said post captain, E. W. Moore, the ship *Austin* and the brig *Wharton*, with their crews, and bring them, or any of them, into the port of Galveston, that the vessels may be secured to the republic, and the culprit or culprits, arraigned and punished by the sentence of a legal tribunal.

The naval powers of Christendom will not permit such a flagrant and unexampled outrage, by a commander of public vessels of war, upon the rights of his nation, and upon his official oath and duty, to pass unrebuked, for such would be to destroy all civil rule and establish a precedent which would jeopardize the commerce of the ocean and render encouragement and sanction to piracy.

In testimony whereof, I have hereunto set my hand and caused the great seal of the Republic to be affixed.

Done at Washington on the 23rd day of March, in the year of our Lord, one thousand eight hundred and forty-three, and of the independence of the Republic, the eighth.

SAM. HOUSTON

[The above proclamation was published in the *Civilian and Galveston City Gazette* on May 10, 1843. Because he had seen James Morgan's copy of the proclamation before they headed off to Mexico, and knew that it might be published at any time while they were at sea, Commodore Moore was prepared to respond in absentia by forwarding the following letter to be published at Galveston. It appeared in the Galveston *Times* May 16, 1843 edition.]

April 19th, 1843
Texas Sloop of War *Austin*,
Outside N. E. Pass of Mississippi

Mr. F. Pincard, Editor of *Texas Times*, Galveston.

In the event of my being declared by proclamation of the president as a pirate, or outlaw, you will please state over my signature that I go down to attack the Mexican squadron, with the *consent* and *full concurrence* of Colonel James Morgan, who is on board this ship as one of the commissioners to carry into effect the secret act of Congress in relation to the navy, and who is going with me, believing that it is the best thing that could be done for the country.

This ship and the brig have excellent men on board, and the officers and men are all eager for the contest. We go to make *one desperate struggle* to turn the tide of *ill luck* that has so long been running against Texas.

You shall hear from me again as soon as possible.

Yours truly,
E. W. MOORE

BRIEF SYNOPSIS

OF THE

DOINGS OF THE

TEXAS NAVY

UNDER THE COMMAND OF

COM. E. W. MOORE;

TOGETHER WITH HIS

CONTROVERSY WITH GEN. SAM HOUSTON, PRESIDENT
OF THE REPUBLIC OF TEXAS; IN WHICH HE
WAS SUSTAINED BY THE CONGRESS OF THAT
COUNTRY THREE DIFFERENT SESSIONS;
BY THE CONVENTION TO FORM A
STATE CONSTITUTION; AND BY
THE STATE LEGISLATURE,
UNANIMOUSLY.

INTRODUCTION

The undersigned regrets that he is compelled to appear before the public in this manner, in vindication of his *sacred honor*, against the charges of murder, embezzlement of the funds of the late Republic of Texas, and of cowardice; which he is, by the *double dealing and duplicity* of General Sam Houston, who, not content with his unjustifiable and illegal course as President of the late Republic of Texas, towards the undersigned, has *secretly* exerted his influence as Senator of the United States, to injure the character and reputation of the undersigned, in the City of Washington, by endeavoring to impress upon persons there that he was forced to fight the Mexican squadron off Campeche in 1843 by the officers under his command; *of which the undersigned has written evidence from a highly respectable source*—and by further causing to be published in that city at the close of the last session of Congress, a pamphlet containing but *three* documents relative to his (the undersigned's) connection with the Navy and late Republic of Texas, vis: "President Houston's Proclamation of Piracy, &c; President Jones' veto message on a bill appropriating about *one sixth* of the pay due the undersigned for services as Captain commanding Texas Navy; and the argument of Thomas Johnson, Judge Advocate" of the special court-martial established by the Congress of Texas, "for the impartial trial of Post Captain E. W. Moore and others." By which tribunal the reputation of the undersigned was fully vindicated; but General Sam Houston prostituted the high office which he filled, as President of a free people, by "disapproving the proceedings of the court in toto," stating, "as he was assured by undoubted evidence of the guilt of the undersigned" although he failed to satisfy the court-martial on that point, every member of which was selected by himself except the President thereof, who was specified in the act creating the special Court Martial. This was done by General Sam Houston only *six or seven days* before his term as President of the Republic of Texas expired in December, 1844.

The main object in the publication of the pamphlet referred to, is the circulation of the Veto Message of President Jones which contains the assertion, backed by the late Auditor of the Treasury of Texas, that the undersigned was a defaulter to a large amount; of which the undersigned has only to state that he had been *previously tried* for this charge of "misapplication and embezzlement of the public funds" *and acquitted*—that he has since obtained a certificate from the *very same* Auditor "that there were no charges against him on the books of his office" upon which he drew money, on the 22nd December 1845, from the Treasury of Texas—and further that since the state authorities of Texas have gone into effect, the undersigned under a law of the Legislature, has obtained a settlement of his accounts by which a large balance *in his favor* will be reported to that body at their next meeting, to be acted on by them in conformity with the law referred to.

The above facts which will be fully shown by the following pages; and the well known fact that General Sam Houston has refused on repeated occasions, to render that redress to which gentlemen are sometimes compelled to resort; are my apology for the publication and circulation of the following pages in vindication of truth and even-handed justice, until I can have published the *entire* records of the Court Martial which have been unavoidably delayed but which will soon appear.

E. W. MOORE,
Captain Commanding
Late Texas Navy.
WASHINGTON CITY, MARCH 27, 1847

The Texas Navy in 1839, when Captain Moore was appointed to the command, consisted of one Brig of eighteen guns, one Steamer of eight guns, and one schooner of seven guns; it was increased that year and the early part of 1840, by one Ship of twenty guns, one Brig of eighteen guns, and two Schooners of seven guns each. His first cruise was in 1839, with the Brig and Steamer—the schooner was left cruising on the coast of Texas to keep in check any cruiser that the Mexican Government might send up to enforce the Blockade on the ports of Texas, which had been declared and published. His second cruise was in 1840 & 1841, with the Ship, three Schooners and Steamer; leaving the two Brigs in Galveston, one of them manned and ready for sea in an hour's notice, to keep in check any Mexican vessel that might slip by him and get off the coast of Texas to enforce the Blockade which had been reiterated—(the Mexican Navy having been increased by a Steamer.) This cruise commended in June, 1840, and the last vessel under his command got into Galveston in April 1841—one Schooner was captured and sent into Galveston, condemned, and sold for over $7,000. Tabasco was taken in November 1840, and a contribution of $25,000 levied and received, with which supplies were obtained from New Orleans to enable the vessels to keep at sea upwards of ten months— one of the Schooners was lost on the Arcas Islands during this cruise—no lives or property lost except the hull of the vessel.

From May to November, 1841, the vessels were overhauled and the coast of Texas surveyed by Captain Moore in the two Schooners; a chart of the entire coast was made by him and published in New York and England, which is the only correct one now in use by Navigators. In the fall of 1841, he sailed again with the Ship and two Schooners, under and arrangement made by President Lamar with an Agent sent to Texas by the authorities of Yucatan, which was sanctioned by the Texas Congress the following session, and is referred to in the report of the joint committee on Captain Moore's memorial to the eighth Congress; (see house journal, page 348 to 361 inclusive.)

After the action of Congress he was joined off the coast of Mexico by *one* of the eighteen-gun Brigs—during this cruise three Schooners were captured and sent into Texas: he returned to Galveston in May 1842 and was sent to New Orleans and Mobile to fit out the Ship, one Brig and two Schooners, to enforce the proclamation of Blockade of the ports of Mexico issued by President Houston in April, 1842—the two Schooners were cruising in the Gulf in early June—they sailed from Mobile—the Ship and Brig were fitting out in New Orleans; the former sprung a leak and had to be docked and repaired—several new spars had to be put in her and the Brig, which exhausted the appropriation of Congress of the preceding winter—the extra session was called at Houston in consequence of the invasion of Texas in March, 1842, appropriated, July 23d, 1842, $15,000 for the repairs and outfits of the Steamer *Zavala*, which vessel was lying in Galveston harbor—$25,000 for the outfit and provisioning of the other vessels of the Navy—and $57,659.50 for the pay of the officers, seamen and marines—not one dollar of which was ever issued from the Treasury.

The Steamer had to be run on the flats in Galveston harbor to prevent her sinking, where she was permitted to lay until the worms destroyed her, when she was broken up and sold in 1843 and 1844—one of the Schooners (the *San Antonio*) foundered in a heavy gale in the gulf in September, 1842, and the other one (the *San Bernard*) was blow ashore in Galveston, where she had arrived a few days previous—she was sent off some time after and thoroughly repaired by the Commissioners—the Ship and Brig remained in New Orleans for the want of funds to fit them out; the government urging Captain Moore to fit them for a cruise and in the fall ordering him to bring them to Galveston if he could not fit them out for a cruise on the Gulf, withholding all the while the appropriation mentioned above—he was not able to do either because he had not men to defend them on their passage to Galveston or even to work them.

The Mexicans in the meantime, that is from May 1842 to the fall of that year, had increased their naval force by two large Steamers built in England, a large Schooner built in New York and the capture of two Brigs and a Schooner from the Yucatacoes—with their whole force of three Steamers, two Brigs and two Schooners, they landed an army of 8,000 troops near Campeche and blockaded that port and Sisal, having previously taken possession of Laguna; thus blockading their entire coast and besieging their principal seaport.

Under this state of things, the Government of Yucatan, knowing the position and condition of the Texas vessels of war in New Orleans, sent the same agent who had been sent to Texas in 1841, and supplied Captain Moore with about $7,000 with which and all the money he had, and supplies he had obtained on credit to the amount of $16,000, he fitted out and manned the Ship and Brig to attack this large force off Campeche, and thereby save Galveston from being destroyed as soon as the people of Yucatan were subjugated. When he was nearly ready for sea, two commissioners arrived from Texas, who had been appointed under a *secret act* of the Congress of Texas—the objects of which Captain Moore was not permitted to know anything about, but was addressed by them as follows:

NEW ORLEANS
Monday, February 27, 1843

SIR:

You will receive herewith a letter from the Honorable the Secretary of War and Marine of the Republic of Texas in regard to the vessels of the Republic under your command in this port; and we should be glad to receive your report with as little delay as practicable.

We have the honor to be
with every respect,
your obedient servants,

J. MORGAN,
WM. BRYAN.

———

DEPARTMENT OF WAR AND MARINE
Washington, 22d January, 1843

To Commander J. T. K. LOTHROP
Or Officer in command of Navy.

SIR:

Immediately upon the reception of the order you will report the condition of the vessels, the number of officers and seamen under your command, to Wm. Bryan, Samuel M. Williams and James Morgan, who have been commissioned by the President to carry into effect a *secret act of Congress with regard to the Navy,* and you will act under and be subject to the order of said commissioners, or any two of them, until you receive further orders from this department.

I have the honor to be
your obedient servant,

G. W. HILL,
Secretary of War and Marine

To which Capt. Moore made the following reply:

MESSRS. J. MORGAN AND WM. BRYAN

NEW ORLEANS
Monday, February 27, 1843
Gentlemen:

Your communication of today, enclosing me one from the Hon. Secretary of War and Marine, dated 22d January, has this moment been received, and as it would consume the whole day to get it on board ship and have muster rolls, &c. made out, I proceed at once to state to you the condition of the vessels and the situation in which I find myself placed, without knowing what is the nature or object of the *secret act of Congress with regard to the Navy*, which the Hon. Secretary of War and Marine has informed me you are commissioned to carry into effect.

The first intimation that I had of the wish of the Department that I should be at the seat of government was on Saturday last (25th inst.) in a letter from the department dated 22d January, which was opened fortunately in the presence of one of you (Mr. J. Morgan.) In the early part of January I received a letter from the department, dated some time in December, in which I was notified that I might not hope even for pecuniary aid from the government to get the vessels to sea, as the appropriation for the Navy, of the extra session of Congress was a "dead letter"—and was informed that if I could by any means get to sea that I was expected to do so and carry out my former orders, which are sealed, to be opened after I discharged the pilot at the mouth of the Mississippi River—upon which I despatched in a few days, a gentleman of this city with a letter for the Governor of Yucatan, to which he very promptly replied, meeting fully my propositions; which were that if the Government would furnish me with the means to get to sea well manned, that I would attack the Mexican Squadron at that time, and now blockading the port of Campeche.

An agent of the Government of Yucatan arrived here on the 7th inst. and on the 11th we entered into a mutual agreement equally advantageous to Texas and Yucatan; and which will I trust, result in the capture of the Mexican Squadron; after which the Government of Yucatan has stipulated to furnish me or my successor with $8,000 per month for the support of the Navy until the termination of her difficulties with Mexico, provided the vessels of the Texas Navy are kept at sea in such force as to prevent a renewal or continuance of the blockade of the ports of Yucatan. All prizes captured belonging to the Government of Texas, with the exception of the first engagement off Campeche in which they stipulate to aid with their gun boats, in which case they share in an equal ratio in the event of their being in the fight, and the four vessels originally belonging to Yucatan, viz: two Brigs and two Schooners (worth very little) are to be returned to them.

In carrying out this agreement, and with the aid of a few friends here who had previously aided me and prevented the vessels being burned, by a number of men who had been discharged without a dollar of their pay, (which they had faithfully earned,) by loaning me upwards of *three thousand dollars* which I divided amongst the men, I have the vessels so far as regards munitions of war, stores and provisions, prepared for a cruise of *three months*, have either on board or engaged to go on board, over half the crews which I hope and expect to complete in a few days, when I will be ready to sail and meet the enemy. We have a full complement of officers who are eager to get to sea and convince the Government what great injustice and neglect they have met at its hands undeservedly.

I have the honor to be
with great respect,
your obedient servant,

E. W. Moore

———

The commissioners then addressed a letter to Capt. Moore (February 28th, 1843,) asking to be placed in possession of the vessels, which Capt. Moore answered (March 2d,) in substance, that he could not do so, because he would then be placed in a position that if he should receive "further orders from the department" as indicated in the letter to him received through the hands of the Commissioners, that he would not be able to comply with such orders, at the same time expressing his readiness to obey any legal order that the Commissioners might see proper to give him. This correspondence was forwarded to the department by the Commissioners, who received in reply the Proclamation, with copies to be sent to the Charges De Affaires of the Republic of Texas, in England and the United States, and a letter *open*, to Capt. Moore, suspending him from command and ordering him to Washington (Texas) under arrest; it being left to the judgment and discretion of the Commissioners or any *one of them*, in the absence of any other, to publish the Proclamation; or in short to act as circumstances might best require for the interest of Texas. The letter of arrest was handed to Capt. Moore by one of the Commissioners, and *immediately afterwards was withdrawn* by him, and Capt. Moore was continued in command, for the preservation of the vessels and the interests of Texas, and in order that they might be carried safely to Galveston under his command. After some delay in getting our full crew we finally sailed on the 15th of April, 1843, for Galveston direct, Commissioner J. Morgan being on board. For reasons which are given by himself under oath, before the special Court Martial ordered by Congress upon the appeal of Capt. Moore to that body, we changed our route, pushed for Campeche, attacked and whipped off the Mexican fleet after a series of skirmishes and fights, which commenced on the 30th April, and terminated on the 26th June, 1843, by the Mexicans running off in the night and leaving Capt. Moore with the two vessels under his command on the spot; for the full particulars of which see the following testimony of Commissioner James Morgan:

EXAMINED BY CAPT. MOORE

Question 3d. Did you or did you not hand me, on or about the 3d April, 1843, the letter from the Department of War and Marine dated March 21st, 1843, as stated in specification 5th of Charge 3d?

Answer. Yes, I handed you the letter named in the question.

Question 4th. Was or was not said letter (dated March 21st, 1843,) sealed when you handed it to me?

Answer. It was not; there was a seal and stamp of the Republic on the envelope, but it was not closed by a seal.

Question 5th. You will please state to the court the purport of the conversation which took place between us upon your handing me said letter, and after I had read the same?

Answer: You looked at it and observed, "The letter was unsealed!" I said, "Yes,"—you remarked, "It was strange the department should send a letter in that manner;" and asked me if I knew the contents of it. I said "Yes, I have a copy of it." We met again soon after, and you remarked to me—"You see the situation I am now placed in: I have been at a vast expense to get these vessels fitted and ready for sea; —you know in what a situation the vessels are now. I entered into a contract with the Yucatan government, which I felt myself legally authorized to do, and am now suspended from command of the vessels by this order. If I obey this order and leave the vessels in the port of New Orleans, every officer under me will resign; the vessels will be left at the mercy of the sailors, who will be sure to mutiny and destroy them; which they threatened to do once before when I could not pay them off."

I said to Commodore Moore, the order the commissioners have now received are painful to me, I assure you. I accepted this commission with a great deal of reluctance, but I mean to execute the orders regardless of consequences. He then asked me if I had a communication for any officer under his command. I told him I had another communication for Captain Lothrop, to take charge of the vessels. He asked me if I had delivered it?—I told him I had put it under a sealed cover and handed it to Mr. Stephens, who was secretary to the commission, to deliver to him; supposing he was on board the brig under his command, which was at anchor below the city.

I think Commodore Moore then repeated what he said to me before of his peculiar situation, the liability of the vessels to be destroyed if he left them—the loss the government might sustain thereby; and remarked to me, if I would withdraw the communication to Captain Lothrop he would proceed with the vessels immediately to Galveston, and go with me direct to the President and endeavor to satisfy him that he never intended to do anything wrong with the vessels, nor anything that would compromise his own honor or that of his adopted country.

I remarked to him "Now, that is precisely what I want done with the vessels. If Captain Lothrop has not received and read the communication I have sent him by Mr. Stephens this evening, the arrangement proposed shall be made, and you shall continue in command of the vessels until our arrival at Galveston." I then sought Mr. Stephens, to know if he had delivered the communication but did not find him, but soon after I met Captain Lothrop and inquired of him if he had received a communication from me sent him by Mr. Stephens. He replied that he had received no communication from me that he knew of. I then remarked to him, "Capt. Lothrop, if you do receive a communication from Mr. Stephens under a sealed cover, it is from me, and I would be glad if you would not break the seal, but return it to me with the seal unbroken." Upon which, he drew from his pocket the said communication, held it up to me and asked me if that was it? I said it was, and that it was a communication from the Department of War and Marine; he then remarked that any communication from the Department of War and Marine to him should have come through Commodore Moore. No further conversation took place between Captain Lothrop and myself.

Soon after, I fell in with Commodore Moore, and remarked to him, "I have got the communication from Captain Lothrop, and he had not broken the seal, and the understanding between us now is that you continue in command of the vessels until we get them into Galveston, where we are to proceed immediately." Commodore Moore replied, "Yes, I pledge you my honor there shall be no delay, and we will go immediately to Galveston." I remarked to him, as it will be saving the Government some expense, I will take passage with you if you have no objection; to which he replied, "I should be happy of your company."

In the morning as soon as I met the other Commissioner, Mr. Bryan, I made known to him the arrangement between Commodore Moore and myself, to know if it bet his approbation; at which he seemed highly gratified and concurred fully therein.

Question 6th. Was it not agreed upon by us that the purport of our conversation and agreement should be put in writing?

Answer. It was.

Question 7th. Did I not address the Commissioners the next day after the letter from the Department of War and Marine, dated March 21st, 1843, was handed to me?

(The accused presented a copy of the letter to the witness, who identified it, and—)

Answer. Yes, you did.

(The accused then entered the same, it being a letter from himself to the Commissioners Morgan and Bryan, dated April 4th, 1843.)

Question 8th. Did you, or did you not communicate to the Government the tenor and purport of your acts, and the arrangement agreed upon between us after you had received the letter from the Department of War and Marine for me, dated March 21st, 1843, which you handed me, as you have already stated?

Answer. We did.

Question 9th. You will please state to the court the reasons of our route being changed after our arrival at the mouth of the Mississippi and all the circumstances connected with the same.

Answer. The next morning after we arrived at the Balize, and came to anchor, there came on board the *Austin* the captains of two vessels and a passenger, who stated that their vessels were just from Campeche; the last vessel, three days and ten hours. On inquiry, we learned from the captain and passenger of the last arrival that the Yucatacoes and Mexicans were about settling their difficulties; that General Ampudia had ceased bombarding the town; that the Governor and authorities of Campeche and General Ampudia were having almost daily interviews, and that it was their intention to unite against Texas; that the division of troops under Barragan and Lemos that went to attack Merida had failed in their enterprise, were defeated, and had capitulated; and by an arrangement agreed upon, the troops were to leave there in twenty days; that General Ampudia had sent the steamer *Montezuma* to Telchac to take the troops on board, to bring them up to Campeche, and that it seemed well understood at Campeche that Ampudia was concentrating all the forces he had in Yucatan, for the purpose of proceeding forthwith to Galveston; and it was believed that he was urging the Yucatacoes daily to join him, inasmuch as Commodore Moore had disappointed and deceived them in the aid promised with the vessels under his command, and that it was the belief of Governor Mendez of Yucatan, with many others of Campeche, that Commodore Moore had been bribed by the Mexican Government not to furnish the aid he had stipulated to do, and that many of the Yucatacoes were getting quite exasperated at Commodore Moore's supposed treachery.

I then made particular inquiry of the gentlemen who furnished this information, of the situation of their steamers, their force and their other vessels of war, and they corroborated the statements I had heard before of them;—that their steamers were badly manned, their chief dependence being upon Englishmen they had on board, who were represented in a state of insubordination, and it was believed that our two vessels could whip their whole fleet if we should fall in with them. The steamer *Montezuma* was at Telchac, alone, *one hundred and fifty miles* to windward of Campeche, waiting for the troops of Barragan and Lemos, and if our two vessels were to proceed to Telchac, we would find her there alone, and might easily capture her; by which means, the balance of the Mexican fleet would easily fall into our hands. The passage to Telchac from Balize might consume about four days, and as there was no doubt of our finding her there alone, as it would be

impossible for the troops that had capitulated near Merida, to reach the coast and embark before we could get
there; and as from the information I had received in New
Orleans about the intention of the Mexicans to make an
attack upon Galveston in the Spring, so fully confirmed
by the information received at the Balize, and from what
General Houston had assured me in the winter, during
the session of Congress, "that there was to be a formidable
invasion of the country; that it was gone and out of his
power to save it; that it would cease to be a Republic in
six months"—all these circumstances taken together, with
information so fully confirmed, that the enemy intended
to make a descent upon the coast, I was induced to hazard
the responsibility of suggesting to Commodore Moore,
to take Telchac and the coast of Yucatan, on our way to
Galveston, "to save the Republic," if I could.

Question 10th. Were you or were you not fully satisfied
that I would have proceeded direct to Galveston from the
Balize, but for your having expressed to me, as stated by
you in reply to the last question, the wish or suggestion
that we should take the coast of Yucatan on our way to
Galveston?

Answer. Yes, I felt assured that Commodore Moore had
not the least intention of going anywhere but Galveston
when I suggested it to him.

Question 11th. You will please state to the Court under
what circumstances, when and where, the Proclamation of
President Houston, dated March 23d, 1843, was shown to
me, as stated in Specification 4th of Charge 4th.

Answer. After Commodore Moore and myself had made
the compromise and entered into the agreement before
stated, about proceeding direct to Galveston with the vessels, a day or two after that, I remarked to Commodore
Moore, "I feel particularly gratified that matters have
taken the turn between us that they have; for if you had

been obstinate or disposed to act incorrectly in any way with the vessels, I had a paper that would have controlled you.—He smiled and asked me what it was? I told him if he would come to my room at any time I would show it to him; the next day after the conversation, he called at my room and asked me to let him see the paper I spoke of—I told him it was a Proclamation from President Houston, which I handed to him and he read it, at which he made the exclamation after reading it, "Good God! Did the President think I was going to run away with the vessels or turn pirate?" or something to that effect. It was understood that what transpired in the room at that time with regard to the Proclamation was strictly confidential.

Question 13th. Did you at any time show me any part of your instructions as Commissioner?

Answer. No, because the commissioners were ordered to keep their instructions secret.

Question 14th. You will please state to the Court the reasons why the Proclamation of President Houston was not published before we sailed from New Orleans?

Answer. Because the commissioners considered there was no necessity for it.

Question 17th. What was the condition and state of discipline of the *Austin* during the time you were on board of her?

Answer. A more perfect state of discipline than was on board of that vessel is beyond my comprehension.

Question 18th. At or about what dates did you go on board the *Austin* and leave her?

Answer. The evening of the 15th of April, 1843, I went on board in the Mississippi—I left her on the 15th of July, 1843, in Galveston.

Question 19th. You will please state to the Court my conduct as Commander of the *Austin* from the time we left New Orleans to our arrival off the coast of Yucatan, and meeting the enemy, and particularly in relation to the mutineers and murderers who had been tried by court martial before we left New Orleans; as for carrying out the sentences of said court martial I am accused of murder?

Answer. The conduct of Commodore Moore, on board of the *Austin* was such as I should have expected from the Commander of a fleet, rather more perfect than I should have expected from a man of his age. He was very particular in his discipline and appeared to understand every discipline necessary on board of a man-of-war—the crew was exercised twice a day at the guns, and instructed in all the arts of naval warfare, that it was necessary for them to know, so far as I could judge.

With regard to the mutineers, the third or fourth day after we were out they were brought forward one at a time, the officers and crew all assembled to hear their sentences read, which were read aloud by Commodore Moore, himself, in presence of all on board. The Laws for the government of the navy were also read.

After the sentence of each one was read, he gave notice to those who were to be executed to prepare themselves for death; that he gave them to the next day to do so. When the officers and crew were all assembled and the prisoners brought up for execution, Commodore Moore read the Rules and Regulations for the government of the navy, and the punishment mutineers were subject to.

He then stated to these who were condemned to be hanged, that they had been fairly tried by a court martial selected for that purpose and had able counsel assigned them to defend them; that after a patient investigation of the whole affair they were condemned to the punishment of death, which he as the officer in command was bound by his oath and the laws of his country to see executed; that it was the first time in his life he had ever to do anything of the sort, and he hoped to God it would be the last. He afterwards made some feeling remarks at which everybody within his hearing was very sensibly affected.

On the next day at twelve o'clock the ship was hove to and the four condemned to be hanged were accordingly hung up at the yard arm till they were dead, when they were taken down, prayers read over each separately and they were cast into the sea.

Question 20th. You will please state to the court my conduct while in sight of the enemy, my efforts to engage them, and my conduct while in battle?

Answer. Commodore Moore was cool and collected when we hove in sight of the enemy, and the enemy were bearing down upon us, at the same time appearing extremely gratified that there was an opportunity of closing with the enemy. While closing with the enemy, as they were bearing down upon us, he ordered his men to keep cool and be deliberate, and take care not to waste their ammunition, and not to fire unless they felt pretty sure their shots would tell. The enemy commenced firing some *twenty or thirty minutes*, I think, before the Commodore suffered a gun to be fired; then he observed to the first lieutenant, (for I was standing close by him,) "They don't intend to let us get any nearer to them; they are paddling off stern foremost, faster than we can come up to them, keep her away a little, so our broadside can bear and, damn them, give it to them!" And then the action became general between the *Austin* and *Wharton*, against the two steamers; the two brigs and two schooners all firing into us at once.

During the whole of the engagement, the Commodore kept cool and collected, and managed his ship with great skill, as I thought. The enemy finally ran off without the reach of our guns. It became perfectly calm, and the action ceased.

While we remained at Campeche, after coming to anchor, Commodore Moore's efforts to engage the enemy were unceasing, which he never could effect until 16th May. Soon after daylight on the 16th, I discovered from the hotel where I boarded on shore, the ship *Austin* and brig *Wharton*, underway standing out of the harbor, and the two Mexican steamers standing out from Lerma, to sea, under a press of steam, and a schooner in company. Soon after sunrise, I could perceive the ship, brig and gunboats all in pursuit of them. About nine or ten o'clock, the action commenced between the ship *Austin*, brig *Wharton* and Yucatan gunboats, and the two Mexican steamers and schooner *Eagle*; the steamers underway under a press of steam, running out to sea; Commodore Moore and his squadron in pursuit of them, and they kept up a running fight for several hours until they were not visible from shore.

Question 22d. You will please state to the Court the reasons of our remaining off the coast of Yucatan so much longer than you anticipated when we left the Balize?

Answer. We were first detained by Commodore Moore waiting to engage the enemy. The steamers and their other war vessels were all in sight while we lay at Campeche, and it was evident if we left there without an engagement and attempt to capture them, that they might carry their original plan into execution about concentrating their troops and making an attack upon Galveston. We were determined to prevent it if we could do it by an engagement with their fleet, which never could be brought about until the 16th of May, which was the last engagement before spoken of.

Question 23d. Did you or did you not report to the Department of War and Marine our movements &c., prior to you knowing that the proclamation of President Houston, in relation to the Navy, was published? If so, please state by what means you sent the report and what was the date of the same?

Answer: As soon after our arrival and the first engagement off Campeche, as we could obtain a vessel to take
despatches to Texas, we did so, which was prior to hearing
the aforesaid proclamation had been published. I communicated fully with the Department of War and Marine and
with the President, of all our proceedings from the time
we left New Orleans up to the sailing of the vessel with
our despatches. The vessel was a small Yucatan gunboat,
which was obtained by Commodore Moore from Governor
Mendez to take our despatches up.

Question 24th. Have you or have you not a copy of the
full report made by you, to the Department of War and
Marine, as stated in reply to the last question? If so, please
present it to the court.

Answer: I have.

(Witness produced a copy, which being compared with a
copy from the Department of War and Marine, presented
by the accused to the court, was found to be correct, and
the latter was received by the court and entered, viz:

CAMPECHE, 9th May, 1843
HON. G. W. HILL, *Secretary of War and Marine*

SIR: When I last wrote to the Department from New Orleans I was on the eve of departing with the *Austin* and
Wharton for Galveston, direct, both vessels having been
completely manned and provisioned; the officers and men
in high spirits, under an impression that the vessels were
to *touch* at Galveston only, and proceed to the coast of
Mexico. For reasons stated in former communications of
the Commissioners, it would not have done to have discharged the crews or to have made known the object and
intention of taking the vessels to Galveston, and I was disposed to have all things go on as quietly as possible until
our arrival. But when we got to the Balize, the information

343

received be several arrivals at that place direct from the Mexican coast, confirming what we had so often heard of the miserable situation of the Mexican Navy on the coast of Yucatan, and above all, that the steamer *Montezuma* was along, *at Telchac, one hundred and fifty miles* from Campeche, and that the mutinous state of the crew was such that our vessels could capture her with ease; this, added to the great anxiety of the officers, now that an opportunity offered of doing something for their adopted country, induced me to consent to take the Mexican coast on our way to Galveston, and capture those steamers or clear the coast of them.

These and other reasons which will be still more satisfactory to the Department, when known, enhanced the inducement to consent to touch at Telchac on our way to Texas, and we accordingly proceeded for that port, where, from light and occasional headwinds, we did not arrive until the *eighth* day from the Balize. As we did not find any of the Mexican Navy there, we proceeded along the coast to Sisal, where we ascertained that the *Montezuma* had passed on her way to Lerma and Campeche near *twenty-four hours.* We accordingly kept along the coast until within some *ten or twelve miles* from Lerma, when we anchored on Saturday night, 29th ult., and on Sunday morning, the 30th, at 4 o'clock got underway. At daylight we discovered two large steamers, two armed brigs and two armed schooners, bearing down evidently to attack us—prepared for action and headed directly for them, the crews of both vessels giving three hearty cheers.

But before we could get within long gun shot range of this formidable fleet, headed by their redoubtable steamers and armed with their renowned Paixhan guns, they evinced a disposition to be off, by going about under a heavy press of steam, and heading directly from us. Commodore Moore crowded sail in pursuit, when the Mexican fleet concentrated and awaited our approach, and so soon as within long range opened their fire upon us from their steamers, to which the Commodore paid no attention, but kept endeavoring to close with them. This, however, the enemy was determined to avoid, their steam enabling them to select their distance; when, the Commodore finding he could not bring them to close quarters, opened his fire from both vessels, and the action became general and lasted for something over an hour.

The steamers hauled off to windward, entirely without reach of our guns, and remained there until it became calm, when they recommenced their attack which continued for nearly half an hour, and they again hauled off, and kept to windward ever since. As for their brigs and schooners, they approached near enough in the first action to get *one broadside* from the Austin, and never came near us afterwards, but have kept at a distance of some *five or ten miles* to windward outside the steamers ever since! In this action the brig *Wharton* lost two men killed and four wounded—the *Austin* had none killed or wounded; but one shot struck her and that doing little or no damage.

The bravery and gallantry of the officers and men of our vessels could not be excelled; as I was an eyewitness, it affords me great pleasure to testify to it. The Commodore gave his orders with great coolness and cautioned the men not to waste their powder, but make every shot tell. The crew (and a gallant one it is, and to whom the whole affair appeared a perfect jubilee,) executed his orders as nearly to the letter as was within their power, and made some of their shot *tell well*, as was evidenced from the employment given to the carpenters of the steamers after the action, in plugging shot holes!

Captain Lothrop managed and fought his brig handsomely, and the only source of regret in the whole affair, with the officers of both vessels, was that we could not close with the enemy. If we had been enabled to have done so, there is not the smallest doubt that we could have carried both steamers in ten minutes! We have understood since the action that the captain of the *Montezuma* steamer and some twenty or more of their men were killed, and a number wounded.

Commodore Moore has never been able to bring them to close action since; they keep to windward out of his reach. We have driven their fleet from Lerma, raised the blockade of Campeche, placed General Ampudia with his besieging army in a very perilous situation, inasmuch as he cannot now communicate with his fleet. His soldiers are deserting and coming into Campeche in considerable numbers daily, and report that independent of their starving condition, they have the *black vomit* among them! He no longer bombards the town, is quiet and would get away, no doubt, if he could. The troops that capitulated near Merida on the 24th ult., and had *twelve days* allowed them to get away, are in a worse situation. Our employment of their squadron has rendered the sending of transports within the limited time impossible; and the division of Barragan is, at this moment, unless entrenched, cut to pieces by Gamboa, or have surrendered as prisoners of war.

Thus by the visit of our two vessels to this coast, we have prevented the Mexican government from subjugating Yucatan and invading Texas with the same forces to have been transported thither by her steamers and transports, and have upset the arrangement of the Mexican government in regard to Texas and Yucatan altogether. As I have said before, we are on our way to Galveston, via the Mexican coast, with the ship and brig, and I hope to be there in a few days. In the meantime, I send this by a fast-sailing pilot boat, furnished by the Yucatan government, to let the Department see that all is straight, so far as regards our movements and will, I hope, meet the approbation of the government.

I have the honor to remain, with every consideration of respect, your most obedient servant,

J. MORGAN
Commissioner, &c.

————

Question 25th. You will please state to the Court, at or about what time and by what means the Proclamation of President Houston, in relation to the Navy, reached us; and what passed between us in relation to the vessels under my command, and the consequences that might ensue to them from that time up to our arrival at Galveston?

Answer. Somewhere about the 1st or 2d of June, we got notice that the Proclamation had been published. The way that we obtained the information that it had been published was through Colonel Cooke and Lieutenant Gray of the *Austin*, who had been up to Sisal in one of the Yucatan gunboats on their way to Telchac. They stated they had seen it in a New Orleans newspaper. I think that Commodore Moore evinced some anxiety lest the crew should get hold of the Proclamation and, seeing themselves outlawed, might mutiny and take possession of the vessels. I urged him to return immediately to Galveston with the vessels, inasmuch as the government did not seem to approbate the course we were pursuing. The Commodore concurred with me in the propriety of returning immediately, and seemed to be greatly chagrined at the idea, after making the sacrifices that he had, of being proscribed by his adopted country as an outlaw. There were some communications then passed between us in writing on the subject of the Proclamation, all of which are before the court.

As soon as the powder arrived that Commodore Moore was waiting for, and he could get his vessels ready for sail-

347

ing, we left for Galveston; previous to which, the steamers and other armed vessels all run off in the night.

We touched at Sisal, for the purpose of communicating with the Governor at Merida, where we remained three or four days until the Commodore had communicated with Governor Barbachano at Merida, when we left for Galveston, where we arrived about the middle of July.

—————

The letters referred to above will be found in the records of the court, which will soon be published—Appendix I, Nos. 114, 115 and 116; and the reports made to the Department by Captain Moore in relation to his fights with the Mexican squadron will be found in the same, Nos. 99, 100, 105, 122 and 123.

The following extracts from the testimony of the Honorable G. W. Hill, Secretary of War and Marine, are inserted as corroborative of the evidence of Commissioner J. Morgan, that Captain Moore was continued in command of the vessels by the Commissioners until their arrival at Galveston.

Examined by Capt. Moore

Question 39th. Did or did not the Commissioners report to the Department of War and Marine under date of April 8th, 1843, that they at that date had in their possession the orders referred to in your reply to last question, and that I was or had been continued in command by them, until the arrival of the vessels at Galveston?

Answer. They state in the report alluded to that they had in their possession the orders alluded to, that the vessels would be taken immediately to Galveston, but did not report that they had continued Commodore Moore in command, but said he would go in one of the vessels and proceed forthwith to Washington (Texas) and report to the Department.

Question 40th. Were you or were you not fully satisfied as Secretary of War and Marine, from the tenor of the aforesaid report of the Commissioners, that I was in command of the vessels at the date, April 8th, 1843, of the Commissioners report referred to in your last answer?

Answer. I cannot say that I was, from the tenor of the Commissioners report. It rather intimated that the senior Commissioner was in charge of the vessels; the report does not state who was in command.

Question 41st. Were you or were you not fully satisfied as Secretary of War and Marine from any of the reports of the Commissioners subsequent to that date, April 8th, 1843, that I was continued in command by them until the vessels arrived at Galveston?

Answer. I was satisfied that you continued in charge of the vessels with their approbation from the report of April 15th, 1843.

Captain Moore returned to Galveston July 14th, 1843, and surrendered himself to the sheriff, who, as will be seen by the accompanying letters, would take no action in the case—but he was received by the citizens with great enthusiasm.

Texas Sloop of War *Austin*
Galveston Bay, July 14, 1843
Mr. H. M. Smythe, *Sheriff of Galveston*

Sir:—Having been proclaimed by the President of Texas an outlaw and a pirate, and all the nations of Christendom in amity with Texas having been called upon to seize and bring me to this port for trial, I herewith inform you that I have voluntarily returned here and surrender myself up to you, for the purpose of meeting the penalties of the law.

I am, very respectfully, your obedient servant,

E. W. Moore,
Com'dg Texas Navy.

Galveston, July 15, 1843

Commodore E. W. Moore, *Commanding Texas Navy*

Sir:—Your note of the 14th inst. is before me, in which you express a desire of voluntarily surrendering yourself to the constituted authorities of the country for the purpose of meeting the penalties of law, provided for in cases of outlawry and piracy, in which, as per the proclamation of the President of the Republic of Texas, you are implicated.

Much as I admire a disposition of submission to the legal tribunals of your country, so manifested, I am not aware, nor do I conceive that it is incumbent on me in my official capacity, to take cognizance of the matter, as I have neither been called upon to do so by the Executive, nor by any judicial tribunal—nor am I in possession of any information that would justify me in such a course.

With much respect, I am your obedient servant,

H. M. Smythe,
Sherrif, G. Cy.

Commodore E. W. Moore

Sir:—The citizens and military of Galveston desirous of giving you and your officers a hearty welcome on your return to Galveston, beg leave to request that your first landing may be at 4 o'clock this afternoon if agreeable to you, to receive this testimony of their friendly feelings, at Menard's wharf.

Very respectfully, your obedient servant,

J. M. ALLEN, *Mayor.*

———

Captain Moore would not leave the *Austin* until the Sheriff was brought on board by the Committee who brought the above letter when he was informed by the Sheriff that his letter would be answered as above.

Captain Moore's two reports to the Department after his arrival will be in Appendix I to the records of the Court Martial, Nos. 122 and 123, and on the 25th July he received from the Department through the Commissioners the outrageous letter of *dishonorable discharge*, which will also appear in Appendix I to the records of the Court Martial, Nos. 125 and 126, and which is in violation of the following joint resolution:

Be it resolved by the Senate and House of Representatives of the Republic of Texas, in Congress assembled, That it shall not hereafter be lawful to deprive any officer in the Military or Naval service of this Republic, for any misconduct of office, of his commission, unless by the sentence of a Court Martial.

DAVID S. KAUFMAN
Speaker of the House of Representatives
ANSON JONES,
President pro tem. of the Senate.

APPROVED, February 4th, 1841.
DAVID G. BURNET

———

The above law is in the 5th Congress, page 145.

Finding nothing said by President Houston in his annual message, (November, 1843) of his extraordinary course and violation of a positive law, Captain Moore appealed to Congress, which was referred to a Joint Committee of both houses. See their report in journals, House of Representatives of 8th Congress, from page 348 to 361, upon which the following Joint Resolution was passed.

Joint Resolution to establish a tribunal for the impartial trial of Post Captain E. W. Moore and others.

SECTION 1. *Be it resolved by the Senate and House of Representatives of the Republic of Texas, in Congress assembled,* That it is due Post Captain E. W. Moore, to have a full, fair and impartial investigation of the charges, upon which an order was issued, dismissing him from the Naval service of the Republic, and as a Court Martial is the proper and legitimate tribunal for such investigation, and as such court cannot be convened, composed of naval officers, it is hereby made the duty of the Secretary of War and Marine to convene as soon as practicable, a court martial, composed of the Major General of the Militia and at least two Brigadier Generals, and other officers, next highest in grade in office, which shall constitute a naval court martial for the investigation of charges against the said Post Captain E. W. Moore, and said court martial shall be governed by all the rules and regulations governing naval courts martial.

SECTION 2. *Be it further resolved,* That the Secretary of War and Marine is required to furnish the said Post Captain E. W. Moore, with a copy of the charges and specifications against him; a copy of which shall also be furnished to said court, when it shall have convened.

SECTION 3. Provides for per diem allowance.

SECTION 4. Extends the provisions of this law to all the other officers dismissed without a trial by court martial.

APPROVED, *February 5th, 1844.*

––––––––––

By the above law it will be seen that Congress recognizes Captain Moore in his rank, *seven months after the date of the letter of dishonorable discharge;* in violation of law, as before stated, and which is dated July 19th, 1843.

The court convened in May, 1844, and adjourned in August, the charges furnished by the Secretary of War and Marine are the following:

Charge 1st. Wilful neglect of duty: with *six* specifications
Charge 2nd. Misapplication of money—Embezzlement of public property and fraud: with *three* specifications.
Charge 3d. Disobedience of orders: with *six* specifications
Charge 4th. Contempt and defiance of the laws and authorities of the country: with *five* specifications
Charge 5th. Treason: with one specification
Charge 6th. Murder: with one specification.

The following is the finding of the court with the action of the President.

The court having fully and maturely investigated the matter submitted to it, in the case of E. W. Moore, Post Captain in Texas Navy, and considered the charges and specifications, the evidence and defence of the accused, proceeded this 21st day of August, 1844, at which time the court had been adjourned from day to day to determine the same—And after such deliberation it is of the opinion that the *six* specifications under the first charge are not sustained. That the three specifications under the second charge are not sustained. That the second, third, fourth and fifth specifications under third charge are sustained, and that the first and sixth specifications under same charge are not sustained. That the five specifications under fourth charge are not sustained. That fifth charge and specification are not sustained. That the sixth charge and specification are not sustained.

The court therefore pronounce it as their opinion, that the accused is guilt of disobedience of orders *in manner and form* as set forth in the specifications second, third, fourth and fifth in charge third.

SIDNEY SHERMAN,
Major General, President.

THOMAS JOHNSON, Judge Advocate

———

The President disapproves the proceedings of the court in toto, as he is assured by undoubted evidence of the guilt of the accused in the case of E. W. Moore, late Commander in the Navy.

SAM HOUSTON
7th December, 1844

At the close of the court martial, Captain Moore received the following letter from the Judge Advocate:

TO COM. E. W. MOORE

SIR—The general court martial created by act of Congress, for the impartial trial of Post Captain E. W. Moore and others, has this day closed its session.

I have the honor to inform you that after full, patient and laborious investigation of the facts and evidence, upon the various charges and specifications preferred against you on the part of the government, by the Hon. G. W. Hill, Secretary of War and Marine, and after mature deliberation had thereon, the court has come to a decision in your case, of the nature and character of which you will be informed by the property authority in due season.

I have the honor to remain, with high respect and consideration, your very obedient humble servant,

THOMAS JOHNSON,
Judge Advocate.

––––––

Captain Moore was never notified of the finding of the court nor the action of the President thereon—and did not see the same until June, 1845, during the extra session of Congress called to act upon the annexation resolutions—when he obtained a sight of the papers by permission of the then Secretary of War and Marine, (Col. W. G. Cooke) who was Captain of Marines on board the *Austin* during the cruise that was denounced as Piratical by President Houston. He did not obtain a certified copy of the records of the court martial until December, 1845.

The Congress could not obtain the records although they were called for by that body; see House Journals, 9th Congress, pages 155, 200 and 224, for the following:

JANUARY 6TH, 1845

Be it resolved, That the Executive be requested to furnish this House with the proceedings of the late Naval Court Martial, in the cases of Post Captain E. W. Moore, and others,—and to inform this House to whom the said proceedings were addressed by the Court.

On the motion of Mr. McLeod, the rule was suspended and the resolution adopted.

EXECUTIVE DEPARTMENT, WASHINGTON

January 9th, 1845

To the honorable the House of Representatives.

The Executive, responding to the resolution adopted on the 6th inst. requesting him to furnish the House with the proceedings of the late Naval Court Martial, in the cases of Post Captain E. W. Moore, and others, and to inform the House to whom said proceedings were addressed, by the court, respectfully transmits the enclosed communication from the acting Secretary of War and Marine, relating to the subject matter of the resolution and containing the information required by the concluding clause.

-ANSON JONES

The letter of the acting Secretary I have not a copy of; it however applied for twelve clerks to copy the proceedings, instead of sending the original as called for the by the House, upon which the following resolution was offered to the House.

To the Committee on Military Affairs, Wm. G. Cooke, chairman, to whom the message of the President, in relation to the proceedings of the Naval Court Martial, with the accompanying documents from the War and Marine Department, reported the following resolution.

Be it resolved by the House of Representatives, That the President be requested to dismiss M. C. Hamilton, Chief Clerk in the Department of War and Marine, and acting Secretary of War, for indecorous language in the House. Laid on the table one day for further consideration.

Captain Moore, was in Washington City, confined to his room by sickness and was thereby prevented reaching Texas until after the adjournment of Congress. These proceedings in Congress were upon a Memorial presented by him, through a member, complaining of injustice having been done him, in withholding his part of an appropriation made by Congress to pay in part the officers of the Texas Navy, and other acts of injustice. To sustain the grounds assumed by him, he referred to the records of the Court Martial in his case.

The House of Representatives re-appropriated his part of the pay due him, as per act of preceding session, but the bill was not acted on by the Senate, as the journals will show. At the extra session of Congress, which was called to act on the Annexation Resolutions in June, 1845, the bill of the preceding session was passed by both houses, and vetoed by President Anson Jones on the last day of the session, in which he re-asserts many of the charges against Captain Moore which the court had acquitted him of— amongst them, those of *murder* and *defalcation,* (the first of which is absurd, and the latter he has relieved himself of again, since Texas has become a state; showing a large balance in his favor.) upon which the following resolutions were adopted.

Senate Journal, Page 75

Saturday, June 28, 1845

Resolved by the Senate, That it is the opinion of this body, that the trial of Post Captain E. W. Moore, under a joint resolution of the Congress of the Republic, approved February 5th, 1844, by the special court-martial convened under said resolution, was final and conclusive.

On motion, the rule was suspended and the resolution adopted.

Resolved by the House of Representatives, That it is the opinion of this body, that the trial of Post Captain E. W. Moore, under a joint resolution of the Congress of the Republic, approved February 5th, 1844, by the special court-martial convened under that resolution, and the finding of said court, fully entitles Post Captain E. W. Moore to continue in his position as commander of the Navy of this Republic.

Resolved, That the thanks of the House of Representatives of the Republic of Texas are justly due Commodore E. W. Moore and those under his command in the service of the Navy of said Republic.

Adopted.

The journals of both houses of the Texas Congress will show throughout that Captain Moore was viewed by them as being still in service, up to the last day of the last Congress, June 28, 1845, when the above resolutions were adopted. Notwithstanding these facts, General Sam Houston caused to be published in the City of Washington, in the month of March, 1847, a pamphlet with this title, "Documents relative to the dismissal of Post Captain E. W. Moore, from the Texian Navy." The principal document in this garbled statement, as mentioned in the introductory remarks, is the Veto Message of President Jones, in which, among other slanders and gross libels, will be found the following:—

The sympathies of Congress and the Executive might indeed be invoked in his (Captain Moore's) individual behalf; but it might well be refused, unless Antonio Landois, James Hudgins, Isaac Allen and William Simpson, who were executed at the yard-arm of the ship *Austin*, by hanging for one hour and until dead, for a similar offense, by order of Captain Moore while under arrest himself, could be restored to life and partake of its efficacy. Could the "deep give up its dead," and the sympathy claimed be made general in its operation, the Executive would gladly listen to its dictates; but he can never sanction a rule which hangs the poor sailor and rewards his officer for offences of congenial character.

The whole circumstances in relation to this matter are simply these—as stated to the convention by Captain Moore, in his appeal to that body, in consequence of the Veto Message of President Jones. On the 11th February, 1842, a mutiny took place on board the schooner-of-war *San Antonio*, lying off the City of New Orleans. The four men named by the President were a part of the crew of that vessel at that time. In the affray, Lieutenant Fuller was shot dead; the two Midshipmen who were on board, Mr. Allen and Mr. Odell, were both shot down; and the Sailing Master, Mr. Dearborn, was knocked down the cabin hatch and the companion drawn over; after which, the crew all escaped to the city, (there being no other officer on board) where the four men named, with several others, were arrested by the authorities at the request of Lieutenant Wm. Seeger, who commanded the vessel and who was on shore at the time.

The vessel sailed in a few days, after shipping a new crew, and joined Captain Moore off the port of Laguna, Yucatan, bringing two of the mutineers, who were as many as Lieutenant Seeger thought it safe to take to sea under the circumstances—he requesting the authorities to keep the others (eleven) in custody until a larger vessel of the Texas Navy arrived to take them on board. Captain Moore immediately ordered a court martial to try the *two*; which he was fully authorized by the laws to do. The court sentenced one of them to be hung, but recommended him to mercy—the other petitioned the court for further time to obtain evidence from New Orleans, which was granted. On the 4th of April, 1842, Captain Moore reported by letter to the Department of War and Marine the fact of his having ordered the court, and received in reply on the 24th April, a letter dated the 14th of the same month in which the following:–

"Your proceedings personally, and courts-martial specially, are approved and the latter confirmed."

Captain Moore arrived at the City of New Orleans with the ship *Austin* about the middle of May, 1842, and in a few days commenced a correspondence for the recovery of the mutineers and murderers, with the authorities first of that city, and afterwards with the governor of the state of Louisiana; which resulted in his being informed by a letter from the governor dated August 27th, 1842, "that upon a special demand being made by the President of Texas upon him, that the mutineers and murderers would be given up."

By the first opportunity after this, Captain Moore wrote to the Department of War and Marine, under date of 7th September, 1842, in which is the following:—

Accompanying are copies of sundry communications in relation to the mutineers and murderers on board the *San Antonio* in February last. In order to obtain them, it will be seen by reference to the letter from the governor of Louisiana of 27th ult., that a special demand must be made by His Excellency the President. I have therefore sent Mr. J. W. Moore over in the *Merchant* for the purpose of procuring that demand, and hope that he will return in her with it, in order that the men may be tried and punished. The authorities here are quite impatient, and if the *Merchant* returns without the demand, the men will be liberated.

He received in reply to the above letter, one from the Department of War and Marine, dated September 15, 1842, in which is the following:—

Your communication of the 7th inst., with the accompanying correspondence had between yourself and His Excellency A. B. Roman, Governor of the state of Louisiana, upon the subject of the detention by the authorities of that state, of sundry prisoners, (Texas seaman,) charged with mutiny on board the schooner of war *San Antonio*, while lying in the Mississippi River, in the month of February last, and the murder of some of the officers of said vessel at the same time and place, have been laid before his Excellency the President, and he has issued the desired demand or requisition upon his Excellency A. B. Roman, which is herewith enclosed.

Upon the delivery of the prisoners to you, or as soon thereafter as the testimony of the witness can be procured, you will order a court martial for their trial. In the prosecution of which, the regulations of the service and the laws of the land will be strictly enforced.

Captain Moore having obtained the demand, the governor consented that the men should remain in prison until he was on the eve of sailing. He received them on board accordingly, and the next day the court martial convened for their trial.

The four men named by the President, and all the others who were apprehended, were tried under the following charges. Murder, and attempt of murder; mutiny; and desertion.

The following communication, signed by every member of the court martial, accompanied the verdicts and record.

TEXAS SLOOP-OF-WAR *AUSTIN*
April 13th, 1843

To Commodore E. W. Moore,

Sir:—We, the President and Members of the Court Martial convened for the trial of Frederick Shepherd and others, have the honor to transmit to you the accompanying documents, being a true record of the evidence and minutes of the court.

In discharge of the painful duty and awful responsibilities imposed on us, we have endeavored to confine ourselves strictly to the law governing courts-martial, and to the evidence that has been brought before us—and we have duly deliberated upon the verdicts returned.

In the trial of Frederick Shepherd, we are of the opinion that there is no evidence before the court to prove that he was aware that a mutiny was to take place, or that he was in a situation to aid or assist in quelling one on the night of its occurrence. We have therefore found the prisoner *not guilty*, and recommend his discharge.

Of the prisoners, Antonio Landois, James Hudgins, Isaac Allen and William Simpson, we have only to say that we deem the evidence, elicited on the trial of each and every one of them, sufficiently clear and distinct, to convict them of each of the various charges and specifications preferred against them, and have therefore sentenced them to death.

We beg to call your attention to the evidence in the case of William Barrington, from which you will find he was deeply engaged in the mutiny on board the *San Antonio*—but it also appears in the evidence that he informed one of the officers that it was to take place.

In consequence of this information, the court have sentenced him to receive one hundred lashes with the Cats.

Of the evidence in the case of John Williams and Edward Keenan, we think it unnecessary to make any comments. Williams, you will discover, is strongly recommended to mercy.

Very respectfully, &c.

Signed by the five officers composing the court martial.

One of the men died in prison, one turned "state's evidence," and one escaped from prison a few days before they were delivered up to Captain Moore.

Acting under the evidence and finding of the court, and the law, which requires the commander of the Navy to carry out the sentences of courts-martial when without the jurisdiction of the country under whose flag he sails, he carried out the finding and sentence of the court literally. In doing which it became his truly painful duty to see that the law was executed in its extreme rigor, on the four men named in the Veto Message. Most willingly would he have avoided it if he could have done so, and not violated his *oath of office*, and what was therefore his imperative duty.

For doing his duty to his country and its laws, Captain Moore has been published to the world by President Sam Houston as a *murderer*, tried by the order of Congress for the country for the same and acquitted; and again re-accused by President Anson Jones of the same heinous crime, and that in a manner and under circumstances, which he will refrain from commenting at this time.

The journals of the convention to form a constitution for the State of Texas will show (page 312) that the following resolution was adopted by that body on the memorial of Capt. Moore, from which the preceding extracts are copied.

Resolved by the Delegates of the people in Convention assembled, That in closing their labors as the representatives of the people, they respectfully recommend to the favorable consideration of the American government, their gallant fellow-citizen, Post Captain E. W. Moore, and request that he may be retained and provided for in the naval service thereof.

This veto of President Jones, re-asserting the charge of defalcation against Captain Moore, having been published in the official organ of the government while the convention was in session, and before the passage of the above resolution, induced him to wish again to meet the charge, and have it finally settled by the courts of the country. By the following papers it will be seen that although suit was *promised* by the Secretary of the Treasury, to be brought against Captain Moore, *he attended the court and no suit was brought.*

CITY OF AUSTIN, *August 29th, 1845*

HON. J. A. GREER, *Secretary of the Treasury*

SIR: The *National Register* of the 21st inst., published at Washington on the Brazos, and which reached here by last mail, contains a statement made by the Auditor relative to my accounts for the disbursement of certain moneys received by me at various times, and which accompanied the veto message of President Jones on a joint resolution passed at the recent extra session of Congress on my behalf. As the above paper is recognized as the government organ, and as the aforesaid statement is false and libellous, I feel it due to myself to request you to adopt some plan, either by suit or otherwise, whereby I may have the opportunity of exonerating myself of this charge of embezzlement of the funds of government, for which I have been once tried and acquitted—and which is held over me for official accusation and publication in the prints of the country, to carry out the purposes of those, who it seems are determined to injure me in the estimation of my countrymen, to their uttermost extent and ability.

I am prepared to meet a suit in any county of the republic that his excellency the President may select, or to submit my vouchers to the arbitration of his cabinet for their action and decision.

Your attention to this before you leave will oblige me.

I have the honor to be, very respectfully,
your obedient servant,

E. W. MOORE

AUSTIN, *August 31st, 1845*

E. W. MOORE, *Post Captain, &c.*

SIR: In answer to your note of the 29th inst., I consider the Auditor an officer of Government entirely independent of the secretary of Treasury, so far as relates to his duties in the settlement of accounts. He is attached to the Treasury Department more as a check than as a dependency—therefore unless there should arise a difficulty between the auditor and comptroller in the adjustment of an account, I as Secretary cannot officially interfere.

I as the head of the Treasury Department intend to take such legal steps as will secure the interests of the republic and bring all unsettled accounts to a close. I shall, as you wish, put your case in suit as soon as possible in the County of Washington.

I am &c.,
J. A. GREER, *Secretary of Treasury*

REPUBLIC OF TEXAS,
County of Washington.   DISTRICT COURT
                          Fall Term 1845

I, J. D. Giddings, clerk of the District Court in and for the county of Washington, do hereby certify that Commodore E. W. Moore called at my office to enquire if a suit had been filed against him by the government of the Republic of Texas, previous to the commencement of the present term of this court, and stated that he would acknowledge service of the process in such suit, if it was then filed, and should it be afterwards filed, he was on his way to the west, but would return some time during the early part of the court, and would then acknowledge the service of such process, and requested me to inform the sheriff of said county that he would acknowledge the process.

And at this time no suit has been filed in this court.

Witness, J. D. Giddings, Clerk of said Court, and the seal thereof, at Brenham, this, the fifteenth day of November, A. D. 1845 (Seal of the Court)

J. D. GIDDINGS
Clerk, D. C., W. County

This precaution was taken by Captain Moore, because it was known that it would be the last court held under the laws of the Republic, and the law required the process to be served *ten days prior* to the time of the meeting of the court, which was the first Monday in November.

On the 22nd December, 1845, Captain Moore *obtained a certificate from the same Auditor who had given the certificate that accompanied President Jones' Veto Message, "that there were no charges against him on the books of his office;" upon which he drew from the Treasury of the*

Republic of Texas on the same day, his per diem attendance before the court martial amounting to one hundred and eighty-four dollars, the same being for attendance ninety-two days.

The following resolutions adopted *unanimously* by the State Legislature, will show fully the wishes of the people of Texas in reference to the officers of her Navy, who protected and defended her sea coast and commerce against fearful odds; and which, coupled with the fact that General Sam Houston's term as Senator, expired on the 4th of March, 1847, will account for his *pretended* support of the bill before Congress which had for its object the incorporation of the Officers of the Texas Navy into the Navy of the United States, to which they are clearly entitled under a fair construction of the joint Resolutions of Annexation, and the solemn pledges made to the government of Texas, by the representative of the United States in that country.

Resolved by the Senate of the State of Texas, That they respectfully recommend to the favorable consideration of the United States government, the officers of the late Texas Navy, who have gallantly defended the interests and flag of the Republic in the Gulf of Mexico, and request that they may be retained and provided for in the naval service thereof.

Senate Chamber
Austin, Texas
May 14th, 1846.

I certify the above to be a true copy of the original resolution unanimously adopted in the Senate of Texas on the 12th inst.

A. C. HORTON, *President of the Senate*
Attest, N. P. Bee, *Secretary of the Senate.*

Resolved by the House of Representatives of the State of Texas, That they recommend to the favorable consideration of the government of the United States, those officers of the late Texas Navy, who so gallantly defended the interests and flag of the Republic on the Gulf of Mexico in the year 1843, and request they may be retained and provided for in the naval service thereof.

House of Representatives
May 13th, 1846.

I certify that the foregoing resolution was adopted unanimously by the House of Representatives of the state of Texas on the 13th day of May, 1846.

S. W. PERKINS, *Speaker of the House*
Attest, Benj. F. Hill, *Ch. Ck. H. R.*

The vessels transferred to the United States Navy by the State of Texas under the annexation resolutions were a ship of twenty guns, two brigs of eighteen guns each, and one schooner of seven guns. The officers who were in charge of them at the time of the consummation of annexation, who had been constantly on duty upwards of six years, and who continued in charge of them at the request of the Agent of the United States government, who received them, until officers arrived at Galveston to relieve them, which was done in June or July 1846, when they were *turned on shore,* without all of them being paid for their service. *Mexicans,* cutting wood for the government steamers on the Rio Grande, or performing any other services for the government, have always been paid for *their* services; while the Texas Naval Officers, who defended the sea coast and protected the commerce of their country, and who had fought and defended their ships gallantly against fearful odds, (with the great advantage of steam on the part of the enemy.) and who thereby lent a strong helping hand to add to this Union an Empire; under the operation of the joint Resolution of the Congress of the United States *proposing* annexation to Texas, have been turned out of those vessels to give place to others, (for whom I have the highest regard and respect,) who have been quietly waiting in a time of profound peace, for their regular promotion in the United States Navy, as it was before the Annexation of Texas with her extensive sea coast. Is this just? Is it generous? Is it right?

E. W. MOORE,
Capt. Commanding Late Texas Navy
WASHINGTON, March 27th, 1847

REPLY

TO THE PAMPHLET

BY COMMANDERS

BUCHANAN, DUPONT, & MAGRUDER

OF THE

UNITED STATES NAVY

ADDRESSED

TO THE

HOUSE OF REPRESENTATIVES,

IN RELATION TO THE OFFICERS OF THE LATE TEXAN NAVY

*To the Honorable Members of the House of
Representatives of the Congress of the United States.*

The undersigned, with the other surviving officers of
the Texan Navy, memorialized your Honorable body in
January last, praying that the existing laws, limiting the
number of officers in the Navy of the United States, might
be so modified as to authorize the President to incorpo-
rate your memorialists into the Naval service of the United
States. The grounds of this application were fully set forth
in the memorial which was referred to the Naval Committee, who reported on it on the 2nd of May last, after having
fully investigated the subject.

The undersigned would have been content to abide the
judgment of your honorable body without uttering another
word upon the subject. But within the last few days, a pamphlet signed by three Commanders in the Navy, (Franklin
Buchanan, S. F. Dupont and George A. Magruder,) remon-
strating the action of the committee, and a (quasi) protest
against any further action in the premises by your honorable body, has been placed in my hands, and as it contains
statements controverting the facts upon which the committee acted, it becomes my duty, not only to myself and
co-memorialists, but to your honorable body, who will be
called upon in the course of your duties to act upon the
application, that I should examine the grounds of the opposition assumed by these officers.

Your memorialists are perfectly willing to accept the issue
presented by these remonstrants. If I succeed in showing
that every *material* statement which they have made, both
as to "law and facts," are wholly erroneous, and frequently
most grossly perverted, it will not be unreasonable to ex-

pect them, as honorable and candid men, to admit their error and withdraw their injurious imputations.

On the other hand, if I fail to accomplish this, I shall be content to admit that any claim to a favorable consideration of our memorial is absolutely groundless.

Passing by any remarks which unnecessarily and unjust personal imputations are calculated to call forth, I proceed at once, and as briefly as possible, to consider the grounds of objection to the action of Congress, as presented in their pamphlet.

With regard to their argument upon the construction of the Joint Resolution of annexation, I will not occupy your time by offering a single observation, since the report of the committee is, in itself, a full and complete answer upon that point. And I suppose it will not be doubted that the opinion of the members of that committee upon construction of law, is at least as worthy of consideration as the opinion of *three* Commanders in the Navy.

I will, however, in this connection, simply remark that if these officers had been so liberal and candid as to have quoted the *whole of the opinion* of the Supreme Court, it would have been seen, that the real ground upon which the mandamus was refused, was want of jurisdiction; and as proof of this fact, I beg to refer to the opinion of the Court in the case of Brashear vs. Mason,6th Howard's Reports, pp. 92, *et passim*.

But whatever may be the *true* interpretation of the Joint Resolutions of annexation, it is perfectly clear, that Texas, one of the parties to this contract of annexation, believed that the terms of the Resolution *did* provide for the officers of her Navy, as is shown by the Joint Resolution of her Legislature, instructing and requesting her senators and representatives to use their influence to have it accomplished. But suppose the opinion of the Supreme Court does go to the extent claimed for it by these officers, what does it prove, simply that the terms of the Joint Resolutions were not sufficiently comprehensive to embrace the object which the parties to the contract designed should be accomplished?—Is it an argument why Congress should not remedy that defect? On the contrary, does it not present the strongest appeal to your magnanimity and justice?— and is there not a moral obligation on the part of Congress to fulfil the expectations of Texas?

After having labored to show that, by the terms of annexation, the officers of the Texan navy "had no claim to admission in the naval service of the United States," these officers seem to have entirely distrusted the force of their reasoning, or at least they seem to think that they have failed to present sufficient ground to deter the action of Congress, for they forthwith proceed to show that, at the time of annexation, "there were no such persons as officers of the Texan navy."

To establish this proposition they rely upon what they call "statements of law and facts," but which I will endeavor to show before I conclude, are in many instances misstatements, frequently perversions, and the deductions always erroneous. And here let me say, that for these officers individually, I entertain a high respect—but I must be permitted to speak of the sweeping and unwarrantable allegations which, I am sorry to say, have been so recklessly made by them, in terms which unfounded and injurious imputations always deserve. If I had been called on to select from the whole Navy, men least accessible to prejudice, I should have named the two officers whose names are first signed to the pamphlet; with the other, I have no acquaintance; and I cannot, therefore, escape the conviction, that they have neglected to examine for themselves, as they should have done, the only sources from which the true "statements of facts and law" could be obtained, but have derived their information from ignorant or prejudiced persons. So wide of the truth are they in every essential particular, and since they have chosen to promulge their views thus obtained in so formal a manner, I think that I do them no injustice when I characterize it as "reckless."

They say that:

The Texan navy, considered as an organized body of *officers, men, ships,* had been dissolved and disbanded long prior to the Joint Resolutions, the ships laid up in ordinary, the officers, for whose welfare the Governor represents the Republic to have been so solicitous, directed to seek in the civil walks of life the rewards to which their services entitle them.

The history of the Texan navy is short, it seems to have been created by the act of January 26th, 1837; so soon as February 5th, 1840, a great part of it was directed to be laid up in ordinary; the residue of the Texas navy dragged out a precarious existence till the secret act of January 16th, 1843, *placed the navy in the hands of commissioners to he sold.*

The execution of that purpose was frustrated by the commanding officer, who contrary to orders, sailed with the navy on the Yucatan expedition, and refused to surrender the vessels to the commissioners under the act. The expedition was disowned by the Republic, and the President by his proclamation of March 23d, 1843, called on all friendly nations to aid the Republic to seize the commander and vessels, and bring them to Galveston, that the "culprit or culprits" might be "arraigned and punished by the sentence of a legal tribunal."

The Republic had manifestly for some time contemplated abandoning the policy, more burdensome than profitable, of maintaining a navy, since the resolution of July 23, 1842, made appropriation for the payment only of what was due to the officers on the 1st of July, 1842, and for *six* months—not a full year—thereafter.

That would bring the establishment down to January, 1843, and it was on the 16th of that month that the act for the *sale of the Navy* was passed.

Subsequently the commanding officer would seem to have surrendered the ships to the proper authorities of Texas, and the Congress, properly considering the Yucatan expedition as the act of the commander for which the subordinate officers and men should not suffer, by an act of February 5, 1844, made a small and inadequate appropriation to be distributed among the officers of the Navy, as *part pay*, for past services, up to December 31, 1843.

Since that date, the most diligent search has failed to find *any appropriation for pay of any officers or men connected with a Navy of Texas.*

The contemplated sale of the ships having been defeated, an act of February 5, 1844, pursues the policy of that of January 16, 1843, and provides for contracting to keep the navy in ordinary. The navy was laid up in ordinary—its officers were treated and considered as disbanded, no further notice of them is taken, nor is any appropriation made for their pay, after the inadequate provision for the *pro rata* distribution, in part payment for *past services* up to December 31, 1843.

It entirely comports with this view, that President Jones, *is understood* to have attempted, after the Joint Resolutions, to fill up the lists of naval officers, and to give new commissions, with a view to the application of the gentlemen so commissioned, for admission, into the navy of the United States. A list of these officers, properly certified, was returned to the Secretary of the navy of the United States, and was the foundation of the proceeding by mandamus before the Circuit Court of the District of Columbia, upon which the opinion of the Supreme Court above referred to, was pronounced.

The plaintiff in that suit, William C. Brashear, was upon that list *as a commander* in the Texan Navy and *it is understood* that the list certified by President Jones differed *materially*, if not entirely, from the names now appended to the memorial.

The list referred to, we are informed, *was* in the Navy Department. It should be there still. Perhaps inquiries at the instance of the Naval Committee might be successful in procuring it.

It would prove, if we are correctly informed, that *none* of the memorialists were *then* considered officers of the Texan Navy.

Now I aver, that every one of these allegations, in spirit, and for the purpose for which they are made are entirely unfounded in fact, and their deductions from the acts of the Texan Congress, to which they refer without quoting, are erroneous, and are not warranted by the terms of the acts themselves, as I shall now proceed to show.

A short synopsis of the operations of the navy subsequently to 5th February, 1840 will show beyond all question that the first assertion is unfounded—so far from the ships being laid up in ordinary at that period, and the officers disbanded, the navy was in complete organization and performing arduous and important services.

In consequence of a proclamation of blockade of the Ports of Texas, having been issued by the President of Mexico in April 1840, the undersigned sailed from Galveston in June, 1840, for the coast of Mexico with the ship

Austin, steamer *Zavala* and schooners *San Bernard, San Antonio* and *San Jacinto*; leaving two brigs in Galveston, one of them manned and ready for sea in an hour's notice, to keep in check any Mexican vessel that might slip by him and get off the coast of Texas, to enforce the blockade. The Texas navy continued off the coast of Mexico, until April, 1841, a portion of the time off Vera Cruz, keeping the Mexican navy in that port—showing themselves off Tampico, Tuspan, Alvarado and the mouth of the Rio Grande—and they went up the River Tabasco, captured that place—levied a contribution of $25,000, with which supplies were obtained from New Orleans to enable the squadron to keep at sea upwards of *ten months*—and thereby kept the Mexican navy from appearing off the coast of Texas to enforce the blockade—we remained in quiet possession of the town of Tabasco for *twenty-one days* and had no shot fired at us as we were leaving. During this cruise one Mexican schooner was captured within *five miles* of Vera Cruz, sent to Galveston, condemned, and sold for over *seven thousand dollars*.

From May to November, 1841, the vessels were overhauled and the coast of Texas surveyed by Captain Moore, with the aid of the officers of *two* of the schooners of the Texas navy; a chart of the entire coast was made by him and published in New York by E. & G. W. Blunt, and in England by the Admiralty. It is the only correct chart now in use by navigators—one of the officers whose name is attached to the published remonstrance to the Hon. House of Representatives has been in service on the Gulf since it was published in 1842; he has doubtless had occasion to use it and I can with confidence call on him to attest its accuracy.

In December, 1841, the undersigned again sailed for the Mexican coast with the ship and two schooners, and he was joined in April off the coast of Mexico by *one* of the 18-gun brigs—during this cruise, three Mexican schooners were captured and sent into Galveston, condemned and sold—the undersigned returned to Galveston in May, 1842, and was ordered to New Orleans and Mobile to fit out the ship, *one* brig and two schooners, to enforce the proclamation of blockade of the ports of Mexico, issued by President Houston in April, 1842—the *two* schooners were cruising in the Gulf early in June—they sailed from Mobile; the ship and brig were fitting out in New Orleans; the former sprung a leak and had to be docked and repaired, several new spars had to be put in her and the brig—these indispensable expenditures exhausted the appropriation of Congress of the preceding winter.

Thus it is most clear that for more than two years, subsequently to the period, when these officers assert that the ships were laid up in ordinary, and the officers disbanded, we find them actively engaged at sea. These facts, with ordinary diligence, were all within reach of every one of these officers, who, in neglecting to inform themselves, have incurred the responsibility of making unfounded and reckless charges.

Where did they obtain the information that the secret act of 16 January 1843, placed the Texan navy into the hands of commissioners to be sold? Have they ever seen that act? It is not to be found in the published laws of Texas. The provisions of that act can only be inferred from the act of 5th February 1844, of which I will speak presently.

I am warranted in saying that it is not true that the execution of the purpose of the commissioners to sell "was frustrated by the commanding officer, who, contrary to orders, sailed with the expedition to Yucatan and refused to surrender the vessels to the commissioners." The testimony of one of the commissioners before the court martial will demonstrate that, so far from this being the fact, I acted under his orders, (which I was ordered to do,) and with his concurrence and advice.

But it is not most extraordinary that these officers assert that the policy of the secret act of 16th January, 1843, was pursued by the act of 5th February, 1844. I said just now that all that was known of the provisions of the secret act of 16th January, 1843, was to be gathered from the act of 5th February, 1844, the fifth section of which is in these words:

Sec. 5th. *Be it enacted, &c.,* That the act approved 16th January, 1843, authorizing the sale of the navy, be, and the same is hereby repealed.

The "policy" of the act of 16th January, 1843, then was to sell the navy; these officers say that the act of 5th February, 1844, pursued this policy. This assertion is made in the very fact of the act, which in the most effectual manner pursues directly the *opposite policy.* Can it be possible that these gentlemen examined this law for themselves? They talk of the reliability of the memorial because we happened to place our construction of the Joint Resolution of annexation within marks of quotation. I am happy they could not find more important instances of inaccuracy. I will show before I have done with their "statement of law and facts," that as flagrant as is this misstatement, it is comparatively insignificant. Equally reckless is their assertion that the Republic had manifestly contemplated abandoning the policy of maintaining a navy, since the "resolution of July 23rd, 1842," and they arrive at this conclusion from the fact that the resolution appropriates money to pay the officers, what was due on 1st July, and for six months thereafter. I ask, if that is not a most impotent, if sincere, conclusion from such premises. Why would not the deduction, that the United States contemplated abandoning the policy of maintaining a navy, from the fact that Congress only appropriates money for its maintaining a navy, from the fact that Congress only appropriates money for its maintenance for one year? But we are not left to any such inferences, the deduction of these gentlemen is Jesuitical and insincere, if they examined the law for themselves, for they have suppressed a portion of it that flatly negatives any such conclusion. The 4th section provides *for the increase of the pay of the officers*, in the following words:

"That the officers in the navy shall hereafter receive the following compensation—to a Post captain, $200 per month, &c.;" the law goes on increasing the pay in every grade. Besides doing this, *the Bill establishes the office of Naval Storekeeper*. Now I do not say that the Republic might not the next week have "abandoned the policy of maintaining a navy," but I do say that they law does not furnish *the remotest* ground to justify any such conclusion in the mind of any candid, unprejudiced and intelligent man. On the contrary, so far as it furnishes any evidence at all of the policy of the Republic, it is *to maintain* the navy, and this conclusion is plain and certain, when taken in connection with the fact that the appropriation for repairs made by the former Congress had been exhausted, and the necessity for a new appropriation was emergent. An invasion had taken place, and a special session of Congress was called for the purpose of providing means of defence, and one of the measures adopted to this end was this very act, appropriated $40,000 or repairs and outfits, independently of the amount appropriated for pay, which these officers declare was an evidence of an intention to destroy the navy. Is not their conclusion just simply a palpable absurdity?

The law referred to will be found on pages 5 and 6 of sixth Congress, special session; in the library of Congress as accessible to these officers as to me, and if they have chosen to take their information from other sources, they are less justifiable in making this "reckless" assertion. If they desired a fair and impartial investigation, why did they not cite the whole law and leave Congress to draw their own conclusion?

When a man undertakes to enlighten Congress upon a subject involving not only a claim of another to certain "privileges," but in a high degree his reputation—justice and fairdealing require that he should give the matter a thorough personal investigation and point out clearly the sources whence he has derived his information, in order that reference might be made to them—this is necessary even where the most candid spirit is evinced in the investigation, for men will honestly differ in opinion as to matters of inference. These officers however pursue no such course. In most cases it is "we have heard," "we understand," or "we have been informed," and on this vague, unreliable foundation they have based a series of allegations and inferences, as a matter of course abounding with errors. I am glad, however, that they have in one single instance afforded me the opportunity to test their inaccuracy—for in the proposition I am now going to consider, they lead *me to infer* that *they* have "failed *after diligent search to find any* appropriation for the pay of officers and men connected with the Navy of Texas after December 31, 1843."

This assertion is connected with another, "that the Congress properly considering the expedition to Yucatan as an act of the Commander, for which the subordinate officers and men should not suffer, by an act of 5th of February, 1844, made a small and inadequate appropriation to be distributed among the officers of the Navy as part pay for past service up to December 31, 1843."

I will dispose of both at once—there is a disingenuousness in this whole paragraph, that is unworthy these gentlemen and when explained I am sure will not subserve their cause, nor do them any credit. They might have known if they had chosen to investigate properly, that Congress not only approved of that expedition, but paid me money accruing from it, and finally passed a vote of thanks for my services. (See Journals of the House of Reps. of 8th Congress, pages 348 to 361—Journal of the House, special session, last Congress, page 86, and Journal of the Senate, same session, page 75.)

I come now to the assertion that the act of 5th of February, 1844, made an inadequate appropriation of pay for 31st December, 1843. The act referred to made an appropriation for pay due up to 31 December 1843 *because that was the close of the year*—on the same day, and it is *on the very next page* of the book—another law was passed appropriating money for the support of the Navy for the year 1844. Now is it possible that these officers could have *made diligent search?* Is it not plain that they have blindly credited the tales of some ignoramus, who was as prejudiced as he was ignorant—look in pages 115 and 116 of the laws of Texas for that year and the laws will be found. This is not all—on 1st February, 1845, a law was passed appropriating money for the expenses of the Navy, thus providing for it up to the last hour of the existence of the Republic—see laws of Texas for that year, page 74.

Now I ask if I deal with these gentlemen in any "spirit of unkindness," if I characterize the manner in which they have chosen to discuss the subject, as disingenuous, and absolutely indefensible upon any ground that should govern and control honorable and upright men? I do not mean to impeach their honor and integrity, for I am sure when they come to find that they have unwittingly lent themselves to the discussion of statements furnished them by some booby, who had not the brains to do it himself, they will be the last persons to justify the pamphlet on which I am commenting.

I come now to the assertion that "President Jones *understood (by whom? by whom?)* "to have attempted after the Joint Resolution, &c., &c."—(please turn back and read this whole assertion, which is fully quoted on page 5). When persons are about to make grave charges, it is but just to the party charged, and becoming to themselves, to base them on a more solid foundation than "*is understood.*" It is not true that "President Jones filled up lists of officers and issued new Commissions, with a view to their application for admission in the Navy of the United States." Now, as they have asserted an affirmative charge, I call upon them to prove it—and until they do, they must be content to endure the odium of making false charges. It is true, that President Jones *did issue one new commission to John G. Tod, and it was notorious beyond all question, that this was done from motives of personal animosity to me.*

The assertion that "a list of officers properly certified was returned to the Secretary of Navy of the United States," is also *untrue.* The list of which they speak was one deposited by Commander Brashear in the Navy Department, *and was not certified at all.* Were the terms "returned to the Navy Department" used purposely to deceive? This is official language, and might be easily construed as meaning a list "returned by the *Texan authorities* to the Navy Department." However, it is hardly worthwhile to criticize inaccuracy of *expression*, in a pamphlet so abounding with inaccurate *statements*—that I doubt whether a greater number could be got in so short a space, by a force equal to hydraulic pressure.

They say they "*were informed* that the list referred to *was* in the Navy Department—it should be there still," &c. Now how easily might they have spoken with certainty upon this point? They were almost daily in the Navy Department, for nearly a week, within this month, and incredible as it may appear, it seems they did not take the trouble to obtain such information as would enable them to speak without *guessing*. If they had enquired at the proper source for information, they would have been spared the mortification of asserting that President Jones had made out, certified and returned a list of officers to the Navy Department, when in point of fact he had done no such thing.

There *is* a list in the Navy Department, the history of which is simply this. Commander Brashear was attached to one of the vessels of the Texan Navy at the time of annexation; he continued on board to take care of the property until it was transferred. He afterwards made application to the Navy Department for pay, for this period, and this *(uncertified)* list was procured by him for the purpose of receiving money due him. The list contained simply the names of those who, like Brashear, had continued on board the vessels, and in charge of public property *after* annexation. This list was deposited for the purpose above stated, and with no earthly reference to the proceedings in the Circuit Court, upon application for a mandamus, as will be seen by an inspection of the record in that case; for, at that time, Commander Brashear had not the *most distant idea* that any difficulty of his incorporation in the United States Navy *could* occur. So confident was he in the correctness of Mr. Donelson's (then Chargè in Texas) conviction that he *was* provided for in the Joint Resolution of annexation.

The list upon which the Committee acted was furnished by me, and was duly certified by the Adjutant General of the State, under his seal of office—this list is dated nearly *two years after* the one above referred to. It contains all the names of *all* the commissioned officers that were on the first, *except Tod*, who was commissioned *subsequently* to annexation, and of course was never confirmed by the Senate of Texas. And the name of every *surviving* officer on that list *is* on the memorial, so that no discrepancy whatsoever exists; all of which these officers might have known, if they had investigated for themselves instead of relying on "we have heard—we have understood—it is understood" or "we have been informed." (See appendix for these lists.)

I will now make a few remarks on the letter of G. W. Hill, Secretary of War and Marine to Lieutenant George C. Bunner, (and not "Banner" as they write it) paraded with such ostentation in their pamphlet, as illustrative of the "notions of Texas of her own obligations"—and that "the officers, for whose welfare the Governor represents the public to have been so solicitous"—were directed "to seek in the civil walks of life the rewards to which their services entitle them."

This letter was originally published by me to show that the writer of it, either believed I had been, or would be captured, by the overwhelming force that I was in sight of when it was penned—it is dated *two days before* we fought the Mexican squadron more than four hours—it was written in reply to a letter from Lieutenant Bunner reporting himself to the Department, but not received by him until the return of the vessels on July 14, 1843, from off Campeche, he having gone down and joined the brig to which he was attached, in a small schooner that I had sent up from Campeche with despatches to the government. The letter was *rescinded* by the Department, and Lt. Bunner was continued in the service—his name is on the list furnished to the naval committee by me as having been confirmed by the Senate of Texas to take rank from June 1, 1840, it is also on the list furnished the Department by Commander Brashear—and it is not on the memorial, for a far better reason than any these officers have offered in their wonderful production; which is simply because he died two years ago.

With regard to their remark, "that the letter of the same Secretary, of July 16th, (19?) 1843, by order of the President, announced the dismissal from the service of the commanding officer for very different reasons." I will simply remark that the circumstances to which that letter refers had been made the subject of investigation by a tribunal expressly created for that purpose by the Congress of Texas, and after a full, elaborate and patient investigation I was acquitted of every charge. It is true that they found that *four* of the six specifications under the charge of disobedience of orders "were proved in manner and form." It appears by the records of the court that the orders were of such a character that they could not be obeyed. All of this, I am sorry to say these officers might have been able to state if they had chosen to inform themselves—and it would have been something to have done so *in a parenthesis*, as they had the *magnanimity* to do in relation to another charge; although the chivalry, which they so arrogantly claim at the close of their pamphlet, should have prompted them to do it prominently, fully and clearly. They are prompted by "no unkind feeling?" they say—this may be so; but it is a distinction without a difference.

They have sped the envenomed shafts furnished them by those, whose malignity outstripped their intelligence, and if they do not take effect, it is from no want of design on their part—a stab may be no less fatal because accompanied by a smile and a protestation of kindness.

I shall dismiss this letter with simply quoting the law of Texas, which rendered any such act of the President a flagrant violation of duty.

Be it resolved by the Senate and House of Representatives of the Republic of Texas in Congress Assembled, That it shall not hereafter be lawful to deprive any officer in the Military or Naval service of this Republic, for any misconduct in office of his commission, unless by the sentence of a court-martial.   (*signed*) DAVID S. KAUFMAN,
Speaker of the House
ANSON JONES,
President Pro. tem of the Senate.
Approved, Feb. 4th, 1841 (*signed*) DAVID G. BURNET

(See page 145, laws of Texas, 6th Congress)

The next Congress passed a Resolution declaring the conduct of the President as unwarranted—and ever after, to the close of the existence of the Republic, recognized me as a Captain in the navy.

I have now done with all the points dwelt upon by these officers, but one—and that I approach with feelings of profound regret. I know not whether Commanders Buchanan, Dupont and Magruder would deserve greater condemnation for reviving these matters, with a full knowledge that they had all been disposed of to my honor, than for failing fully to inform themselves upon all the facts of the case.

I will demonstrate before I have done, that these three gentlemen are perhaps the persons, of all others in the United States, who were bound by every principle of consistency, justice and every consideration that governs high minded men—not to condemn me in the matters to which I refer, but to accord to me their highest praise for the performance of what I will bring home to them, they consider to have been an act of stern and imperative duty.

The matter to which I refer is the execution of four men under my command, under circumstances which I will now briefly relate.

On the 11th of February, 1842, a mutiny took place on board the schooner of war *San Antonio*, laying off the city of New Orleans; the four men named by President Jones, in the extract quoted from his veto, were a part of the crew of that vessel at the time. In the affray, Lieutenant Charles Fuller was shot dead. The two Midshipmen on board, Mr. Allen and Mr. Odell, were both shot down, and the sailing master, Mr. Dearborn, was knocked down the cabin hatch and the companion drawn over; after which the crew all escaped to the city (there being no other officer on board) where the four men named, with several others, were arrested by the authorities at the request of Lt. Wm. Seeger, who commanded the vessel, and who was on shore at the time.

The vessel sailed in a few days, and joined the undersigned off the port of Laguna, Yucatan, *bringing two of the mutineers*, who were as many as Lt. Seeger thought it safe to take to sea under the circumstances—he requesting the city authorities to keep the others, eleven in number, in custody until a larger vessel of the Texas navy arrived to take them on board—the undersigned immediately ordered a court martial to try the *two*, which he was fully authorized by the laws to do. The court sentenced one of them to be hung, but recommended him to mercy;—the other petitioned the court for further time, to obtain evidence from New Orleans, which was granted. On the 4th of April, 1842, the undersigned reported by letter to the Department of War and Marine, the fact of his having ordered the court, and received in reply on the 24th of April, a letter dated the 14th of the same month, in which was the following: "Your proceedings personally, and of courts-martial especially, are approved and the latter confirmed."

The undersigned arrived at the city of New Orleans with the ship *Austin*, about the middle of May, 1842, and in a few days commenced a correspondence for the recovery of the mutineers and murderers, with the authorities, first of the city and afterwards with the Governor of the State of Louisiana, which resulted in the undersigned being informed by a letter from the Governor, dated the 27th of August, 1842, "That upon a special demand being made by the President of Texas upon him, the *mutineers and murderers* would be given up."

The government of Texas was duly informed by me of the determination of the Governor of Louisiana, and in reply, I received the following letter:

Your communication of the 7th instant, with the accompanying correspondence had between yourself and his excellency, A. B. Roman, Governor of the State of Louisiana, upon the subject of the detention by the authorities of that State, of sundry prisoners, (Texas seamen,) charged with mutiny on board the schooner of war *San Antonio*, while lying in the Mississippi River in the month of February last, and the murder of some of the officers of said vessel, at the same time and place, have been laid before his excellency, the President, and he has issued the desired demand, or requisition, upon his excellency, A. B. Roman, which is herewith enclosed.

Upon the delivery of the prisoners to you, or as soon thereafter as the testimony of the witnesses can be procured, you will order a court-martial for their trial, in the prosecution of which, the regulations of the service and the laws of the land will be strictly enforced.

I wish to call attention here to the fact that, besides the general authority I possessed under the law to convene a court martial, I had an imperative order to do so from the Department.

Although I am desirous to make this as brief as possible, I do not think I am at liberty to omit the letter of the court accompanying their proceedings.

Texas Sloop-of-War *Austin*
April 13th, 1843

To Commodore E. W. Moore,

Sir:—We, the President and Members of the Court Martial convened for the trial of Frederick Shepherd and others, have the honor to transmit to you the accompanying documents, being a true record of the evidence and minutes of the court.

In discharge of the painful duty and awful responsibilities imposed on us, we have endeavored to confine ourselves strictly to the law governing courts-martial, and to the evidence that has been brought before us—and we have duly deliberated upon the verdicts returned.

In the trial of Frederick Shepherd, we are of the opinion that there is no evidence before the court to prove that he was aware that a mutiny was to take place, or that he was in a situation to aid or assist in quelling one on the night of its occurrence. We have therefore found the prisoner *not guilty*, and recommend his discharge.

Of the prisoners, Antonio Landois, James Hudgins, Isaac Allen and William Simpson, we have only to say that we deem the evidence, elicited on the trial of each and every one of them, sufficiently clear and distinct, to convict them of each of the various charges and specifications preferred against them, and have therefore sentenced them to death.

We beg to call your attention to the evidence in the case of William Barrington, from which you will find he was deeply engaged in the mutiny on board the *San Antonio*—but it also appears in the evidence that he informed one of the officers that it was to take place. In consequence of this information, the court have sentenced him to receive one hundred lashes with the Cats.

Of the evidence in the case of John Williams and Edward Keenan, we think it unnecessary to make any comments. Williams, you will discover, is strongly recommended to mercy.

Very respectfully, &c.

Signed by the five officers composing the court martial.

Now, I ask, what course under these circumstances, would any one of these officers have pursued? I did what I *knew* to be my duty. Unpleasant as it was I did not hesitate a moment; I carried out the law by the execution of the offenders.

Do these officers really believe that in doing this, I am guilty of anything for which I deserve to be dragged to the bar of public opinion for censure? They *dare* not think so—for the sake of their own honor, integrity and consistence, they dare not. The fearful circumstances that gave rise to the execution of three men on board the United States brig *Somers*, is fresh in the recollection of everyone. God forbid that I should mention it here, for the purpose of passing an opinion as to its propriety or impropriety, or even to imply a censure. I do it for no such purpose. These officers, I doubt not, see my purpose at a glance—they approved the conduct of that commander. One of them, I know, was his most ardent and able defender, and stood by him with an unquailing front, amidst the terrific tempest of public opinion which the circumstance elicited. Now, in that case, the execution was preceded by no forms of law—no regularly authorised tribunal was organized to hear and weigh with calm and impartial deliberation the testimony for and against the prisoners—they were not defended by counsellors learned in the law. The execution, if justified at all, must be under the operation of that stern, uncompromising necessity, that overrides all law. A proper tribunal acquitted him, so it did me; and the Congress of Texas moreover passed me a vote of thanks. Now then, I call upon these gentlemen, in view of their opinions and acts in relation to the case to which I have referred, to reconcile it with manly generosity—their taking up this fetid carrion of a charge.

I may as well, in this place, introduce the action of the Congress of Texas on the veto message of President Jones,

391

which they have spread before Congress with so much parade. Immediately upon the reading of the message, the following resolution was adopted under a suspension of the rules:

Resolved by the Senate, That it is the opinion of this body that the trial of Post Captain E. W. Moore, under a joint resolution of the Congress of the Republic, approved February 5th, 1844, by the court-martial convened under said resolution, was final and conclusive.

And again—House Journal, page 88, same date:

Resolved by the House of Representatives, That it is the opinion of this body, that the trial of Post Captain E. W. Moore, under the joint resolution approved February 5th, 1844, by the special court-martial convened under that resolution, and the finding of said court, fully entitles Post Captain E. W. Moore to continue in his position as commander of the navy of this Republic.

Resolved, That the thanks of the House of Representatives of Texas are justly due Commodore E. W. Moore and those under his command in the service of the navy of said Republic.

I will now add in my final disposition of this veto—*That the money appropriated by the act, I have long since drawn from the Treasury of Texas.*

I have as briefly as possible examined the "statement of law and facts" contained in the extraordinary pamphlet issued by these officers, and upon which they rely to defeat the favorable action of Congress upon our memorial, and I think I have shown every material statement, both of "law and facts," is erroneous.

They say that they have "made some of the statements with pain and reluctance, not with any unkind feeling;"

they would have given more substantial evidence of this if, in their pamphlet, instead of fathering the random assertions of prejudiced ignorance and the ebullitions of personal animosity, they had evinced a desire to obtain authentic and undoubted information on points which they have most unjustifiably endeavored to use to my injury.

They talk of "invading their privileges," as if they were "Lords of the Manor," and had detected us in poaching upon their warren. One would imagine, that so far from the navy being a great institution, created for the public good, that it was established for their especial benefit, and was a part of their private estate. Privileges! Why, strictly speaking, as officers, they have not one—except that of resignation; they are creatures, (officially,) into whose nostrils Congress has breathed the breath of life. Promotion is no privilege, it is a prerogative of the power that governs them—and it is this very notion, that they possess some inherent right, that now weighs like an incubus on the navy, and makes the "melancholy privilege" they talk of a criterion to promotion, instead of merit and usefulness to the public. It is not the individual who is employed, but it is his brains, his capacity for performance of services for the public good; and in proportion as he possesses these qualifications, and exercises them in furtherance of the public service, he is *rewarded* by promotion to higher status—and certainly it should not be otherwise.

But they undertake to read a lecture to Congress as to their duty; they inform your honorable body "that you should not except upon the plainest and most obligatory stipulations of treaties of the United States, *for a moment,*" &c., and that too upon *their* construction of a treaty, for they do not leave Congress that "melancholy privilege." I know not what may be the consequences of contumacy of the Committee on Naval Affairs, (to say nothing of the Governor and Legislature of Texas,) for they have been imprudent enough, (the opinion of these officers to the contrary notwithstanding,) to assert that they think a treaty stipulation *does require* that our prayer should be granted.

All that is said in their pamphlet about rewarding "a roaming and unsettled spirit of adventure" is just inane, senseless talk, and only calls forth the reply, that if there be any truth in what has been for a few past years, the frequent subject of newspaper notice and censure, it would vastly conduce to the public interests, and to the efficiency of our navy, if a *less* disposition to *remain settled* and *keep from roaming* were evinced by many of its officers, though I take pleasure in saying that this remark cannot, and is not meant to apply to these three officers, whose professional services and worth are beyond question. All such remarks, however, when applied to the officers of the late Texan navy are, therefore, simply silly and irrelative to the matter at issue; which merely and entirely is, whether, under all circumstances connected with the annexation of Texas, and as a matter of justice to that State, the officers of her navy should, or should not be incorporated into the navy of the United States, without the slightest reference to individuals.

Now as to the "long and fierce war" that was waged with Mexico in defence of Texas, and the "achievements" so ostentatiously referred to—Vera Cruz, Tuspan, Tampico, Tabasco and Alvarado—I beg their pardon: Alvarado was somehow most unaccountably omitted from the list. I am not disposed to detract from the efficiency and activity of the navy in that war. It did, no doubt, all that could be done under the circumstances by any navy. All this and the "Empire which outshines the wealth of Ormus or of Ind," and the conquest of which is such a gem in the naval crown, has really about as much to do with the question as it has to do with the orbit of Uranus.

I would be glad to know if the acquisition of California is to weigh in one scale, why may not the acquisition of Texas be placed in the other. The act of annexation estops any citizen of the United States from disparaging the *value* of the acquisition. The *nation* knew all the risks involved in that act and was willing to encounter them. Besides, it is to Texas these officers owe all the credit of gaining that gem of which they boast; but for the annexation of Texas, California would not have been acquired, and they would have been deprived of the "melancholy privilege" of making a rhetorical display. So that, after all, according to their own reasoning, we, the memorialists, are entitled to their thanks and to the favorable action of Congress; for, as far as our humble ability went, we assisted in acquiring Texas, and thus, indirectly California, and by consequence, the large capital glory of which they are so lavish.

One word on their closing paragraph of misstatements, and I have done. That paragraph is an epitome of the style and manner of the whole pamphlet. The grossest errors are enunciated with a degree of cool confidence and precision, well calculated to mislead and give to their *airy nothings* a local habitation and a name. I will quote the whole paragraph:

> It is then unreasonable for us to complain that these gentlemen having left that country and its flag to seek fame elsewhere, should now mount over the shoulders of their seniors and former superiors in command— stop up the crowded avenues of promotion—and reap the rewards of services they did not perform and dangers they did not encounter.

Now would not any reader suppose that all "these gentlemen" (the memorialists,) had been in the service of the United States and had left it? But so far from this being the fact, I am the *only one* of the whole number who has ever been in the navy of the United States. The rest, like Steuart, Hull and other *heroes* of the American navy, who walked the "avenue of promotion" by the right of merit and not of "melancholy privilege," came into the Texan navy from the commercial marine.

So far then as the other memorialists are concerned, the objection falls; but what, I would ask, has our past history and occupation, which has been reputable and honest, to do with the question at issue. With equal force and propriety, as valid an objection could be argued against our sizes, the color of our hair, or any other physical peculiarity. Again, the prayer of the memorialists, if granted, does *not* "stop up the crowded avenues of promotion," as these officers would have clearly perceived, had their blinding prejudices permitted them to look at the proposed bill, which provides for the *increase* of the personnel of the navy, to the small *extent only* of the number of memorialists, and provides also, that their places shall remain vacant, when they become so by death, resignation, &c.

All of which is respectfully submitted in behalf of the other memorialists and himself, by your obedient servant,

> E. W. MOORE,
> *Capt. Com'dg late Texas Navy*
> Washington, July 31, 1850

APPENDIX

WASHINGTON CITY, D. C.
July 17th, 1850
HON. WM. B. PRESTON
Secretary of the Navy

SIR: You will oblige me by causing to be furnished me a "list of the officers of the late Texas navy," which I am informed has been sent to your Department, and which differs from the one furnished me by the Adjutant General of the State of Texas, in whose office the records of the Department of War and Marine of the Republic of Texas were placed by act of the Legislature of that State.

The Naval Committee of the honorable House of Representatives acted on the authentic list furnished by me, in making their report, accompanied by a bill, having for its object the incorporation of the officers of the late Texas navy, into the navy of the United States; and it will, doubtless, be necessary that I should explain the difference in the two lists.

> I have the honor to be, very respectfully,
> your obedient servant,
> E. W. MOORE,
> *Captain Commanding late Texas Navy*

NAVY DEPARTMENT, July 17, 1850

SIR: In compliance with your request of this date, I enclose, herewith a "list of the officers of the late Texas navy," as prepared by Wm. G. Cooke, dated Department of War and Marine, Austin, January 26, 1846.

> I am, very respectfully, your ob't servt.
> WM. BALLARD PRESTON

(copy)

DEPARTMENT OF WAR AND MARINE,
Austin, January 26, 1846

WM. C. BRASHEAR,
Commander, T. N.

SIR: In compliance with the request contained in your letter of the 16th inst., you are herewith furnished a list of the officers of the navy, at present in service, with their rank and date of their appointment.

I have the honor to be, &c.,
W. G. COOKE

John G. Tod, Captain, commission bearing date June 23, 1840 (*see appended letter of Adjutant Gen. C. L. Mann)

Wm. C. Brashear, Commander, appointed Sept. 23, 1844

A. J. Lewis, Lieutenant, January 1, 1840

George C. Bunner, Lieutenant, June 1, 1840

Wm. A. Tennison, Lieutenant, July 19, 1840

Norman Hurd, Purser, January 16, 1839

J. F. Stephens, Purser, September 21, 1841

C. J. Faysoux, Midshipman, June 29, 1842

H. S. Garlick, Midshipman

––––––––

Compare the above list with the following made out by the same officer, showing all the officers confirmed by the Hon. Senate of Texas, and not those "in service," that is on duty at a given date—and it will be seen that the name John G. Tod is not on it, although other officers are on it, who were appointed the same year, and one of them in the same month.

List of Naval Officers whose appointments were confirmed
by the Hon. Senate, 20th July, 1842

NAME	RANK	No.	DATE	Remarks by E. W. Moore
Edwin W. Moore	Post capt., commanding	-	Apr. 21, 1839	
J. T. K. Lothrop	Commander	-	July 10, 1839	Dead
D. H. Crisp	Lieutenant	-	Nov. 10, 1839	Dead
Wm. C. Brashear	Lieut.	1	Jan. 10, 1840	Dead
Wm. Seeger	Lieut.	2	Jan. 10, 1840	Dead
Alfred G. Gray	Lieut.	3	Jan. 10, 1840	
A. J. Lewis	Lieut.	4	Jan. 10, 1840	
J. P. Lansing	Lieut.	5	Jan. 10, 1840	Dead
Geo. C. Bunner	Lieut.	-	June 10, 1840	Dead
A. A. Waite	Lieut.	1	Sept. 10, 1840	Dead
Wm. A. Tennison	Lieut.	2	Sept. 10, 1840	
Wm. Oliver	Lieut.	3	Sept. 10, 1840	
Cyrus Cummings	Lieut.	4	Sept. 10, 1840	
C. B. Snow	Lieut.	-	Mar. 10, 1842	
D. C. Wilbur	Lieut.	-	June 1, 1842	Dead
M. H. Dearborn	Lieut.	-	July 1, 1842	Dead
R. M. Clarke	Surgeon	-	Nov. 22, 1840	Dead
Thos. P. Anderson	Surgeon	-	Sept. 10, 1841	Dead
J. B. Gardner	Surgeon	-	July 20, 1842	
Norman Hurd	Purser	-	Jan. 16, 1839	
F. T. Wells	Purser	-	June 10, 1839	Dead
J. F. Stephens	Purser	-	Sept. 21, 1841	
W. T. Brannum	Purser	-	July 21, 1842	Dead
Jas. W. Moore*	Purser	-		

*appointed in place of W. T. Brannum

N. B.—The lowest number of the same date takes rank

Department of War and Marine,
July 21, 1842

Officers enumerated in the above list will take rank in the order in which their names are placed.

(signed)   GEO. W. HOCKLEY,

Secretary of War and Marine

State of Texas,
Adjutant General's Office

I, William Gordon Cooke, Adjutant General of the State of Texas, do hereby certify, that the foregoing is a correct copy, taken from the records of the Department of War and Marine of the late Republic of Texas; which records have been attached to this office by act of the Legislature of the State of Texas.

Given under my hand and official seal, this 30th day of November, A. D. one thousand eight hundred and forty-seven.

WM. G. COOKE,
Adjutant General

No other naval appointments than those in the preceding list *were ever confirmed by the honorable Senate of Texas.*

The following letter from the Adjutant General of the State of Texas, will, I trust, satisfy even these officers, of the validity of the claims of the *person* whom they say "claims the dignity for which E. W. Moore is now contending."

Sam Houston vs E. W. Moore

To Commodore E. W. Moore, Austin, Texas

SIR: Yours of the 7th instant has been received. You request me at my earliest convenience to inform you of the date of the commission as captain, Texas navy, given to John G. Tod, by the last President of Texas, and the time said commission was given to him. Also, whether it appears in the records of my office, that said John G. Tod was nominated to the honorable Senate of Texas, and confirmed as a "captain in the Texas navy."

In reply to your first enquiry, as to "the date of the commission as captain, Texas navy, given to John G. Tod, by the last President of Texas," I have the honor to inform you that I find from the records of the Department of War and Marine, now in my office, that John G. Tod was commissioned on the *27th day of Aug.*, 1845, (two months after the last Congress of Texas had adjourned, and fifty-one days after the convention to form a State Constitution had agreed to the terms of annexation, and were deliberating with the United States flag flying over the capitol,) *to rank as captain from the 23rd of June*, 1840. As to your second enquiry, I have compared the certified list of naval officers whose appointments were confirmed by the honorable Senate, 20th July, 1842, (given by William G. Cooke, late Adjutant General of the State of Texas,) and find that it corresponds with the record of the Department of War and Marine of the late Republic of Texas, which has been attached to my office. I have carefully examined the records of the Department of War and Marine, and assure you that there is no evidence of the nomination of John G. Tod to the honorable Senate, in the Department; and as such nomination was required by law to be made to the honorable Senate for the confirmation by the President; in all cases then, where the President failed to make such nominations, and made appointments independent of that *body, who alone* had the right to *confirm* such appointments, made contrary to the provisions of the law, would be a nullity, and his commission a blank.

Agreeably to your wish, I herewith return the certified list of "naval officers whose appointments have been confirmed by the honorable Senate," also a copy of your letter.

I have the honor to be, &c.,

(signed)   C. L. MANN
Adjutant General, State of Texas

REPLY TO THE SECOND PAMPHLET

BY COMMANDERS

BUCHANAN, DUPONT, & MAGRUDER

OF

THE UNITED STATES NAVY

ADDRESSED

TO THE HOUSE OF REPRESENTATIVES,

OBJECTING

TO THE INCORPORATION OF THE OFFICERS OF THE LATE TEXAS NAVY IN THE NAVY OF THE UNITED STATES

*To the Senate and House of Representatives
of the United States:*

The second clause of the second section of the Joint Resolutions of Annexation provides that "said State, when admitted into the Union, after ceding to the United States all public edifices, fortifications, barracks, ports and harbors, navy and navy yards, docks, magazines, arms, armaments, and all other property and means pertaining to the public defence, belonging to said Republic of Texas, shall retain all the public funds, debts, taxes and dues of every other kind, which my belong to, or be due and owing to said Republic." (5th vol. Statutes at Large, p. 797.)

It will be perceived that the navy is here ceded by name and is contradistinguished from all other property and means pertaining to the public defence. The question arises, what is a navy? If this government were to agree by treaty to furnish a navy to France, of twenty ships of the line, to be used in a war with England, would it comply with the terms of such a treaty by furnishing twenty naked ships, without officers or men? If this government were to stipulate to furnish an army of 20,000 would it comply with its obligations by furnishing the soldiers without officers to command them?

Webster defines the navy to be the whole of ships-of-war belonging to a nation or king. The *navy* of Great Britain is the defence of the kingdom and its commerce. This is the usual acceptation of the word. The officers and men "belonging to a navy."

Falconer defines the navy thus: "Navy implies, in general, any fleet or assembly of ships. It is, however, more particularly understood to be the fleet of vessels of war that belong to a prince or state."

"The royal navy of England has been considered its greatest defence and ornament; the floating bulwarks and 'wooden walls' of the island have, therefore been assiduously cultivated and improved from the earliest ages."

"The navy had formerly been victualled by contract, but the victualling is now under the commissioners..."

"NAVY denotes, also, the collective body of officers and men employed in his Majesty's sea-service."

It is too plain for argument, that *mere ships* cannot constitute a navy. If it had been intended to acquire nothing but the vessels or ships, the resolutions of annexation would have used the words "national vessels" or "vessels of war."

Mr. Calhoun argued in the Senate that there was an implied faith which should induce them (the Senators) to pass the bill; and that the *personnel* of the Navy of Texas was to be considered as part and parcel of the same. He denied that there was any overslaughing in the matter; that the bill created as many vacancies as there were officers in the Texas Navy. (*National Intelligencer*, Friday, July 31, 1846.)

Second. There were actually transferred and delivered to the United States *one* ship of *twenty* guns, *two* brigs of *eighteen* guns each, and *one* schooner of *seven* guns.

At the time of the acceptance of the resolutions of annexation by the Convention of Texas, those vessels were the Navy of Texas, and the officers now applying were in commission as the officers of the same; and in May, 1846, when the transfer of the forts, docks, magazines, navy, &c., was actually made, in compliance with the resolutions of annexation, the flag of Texas was hauled down from their peaks and that of the United States hoisted in its stead. The officers and men in charge of the vessels continued on board in charge of them for nearly *three months*, when they were relieved by officers of the United States Navy, who were ordered to Galveston by the Secretary of the

Navy; they were informed "that their services were no longer required," and were put on shore. The petty officers and men continued in the service of the United States for the periods for which they had enlisted in the service of Texas and were taken to Pensacola in the ship *Austin*.

Third. It has been objected to the incorporation of the officers of the Texas Navy into the navy of the United States; that Captain Moore was dismissed therefrom by order of the President of the Republic on 19th July, 1843. The answer to which is twofold: *first*, that it does not affect the general question nor the rights of the other officers; *second*, that it is not true in point of fact, because the President had no more jurisdiction under the laws of Texas to dismiss him than Captain Moore had to dismiss the President of the Republic of Texas. The pretended act was a mere *nullity*.

The law of the Texas Congress on that subject is as follows:

> *Be it resolved by the Senate and House of Representatives of the Republic of Texas in Congress assembled,* That it shall not hereafter be lawful to deprive an officer in the military or naval service of the Republic, for any misconduct in office, of his commission, unless by the sentence of a court-martial.

Approved, Feb. 4, 1842
(*See page 145, Laws of Texas, 6th Congress.*)

Fourth. It has been urged in a pamphlet published by Three Commanders in the Navy of the United States, who have made themselves busy in misrepresenting the principal facts connected with this subject, that the Texas Navy did not exist as an incorporated body at the period of annexation. This position is sufficiently refuted by the facts already stated, and positively disproved by the following acts of the Texas Congress:

Be it enacted by the Senate and House of Representatives of the Republic of Texas in Congress assembled, That the naval establishment of this Republic shall be composed of *one* Captain, *one* Master Commander, *eight* Lieutenants, *ten* Midshipmen, with such other warrant and petty officers as may be necessary for the establishment, upon the scale provided by this act, with sixty seamen and marines, and *one* Lieutenant of Marine, *one* Surgeon, *one* Assistant Surgeon and *two* Pursers.

Approved, Jan. 18, 1841

(See Laws of Texas, 5th Congress, page 109, sec. 9)

For pay of seamen, officers and marines in the Navy, including pay due for the year 1840, *fifty thousand dollars.*

Approved, Feb. 3, 1841

(Same as above, page 110.)

For pay of officers and seamen of the Navy, and for support of the Navy, according to the act to fix the Navy establishment of the Republic, approved January 18, 1841, *twenty thousand dollars.*

Approved, Feb. 3, 1842

(See Laws of Texas, 6th Congress, page 98.)

SEC. 1. That the sum of *fifteen thousand dollars* be, and the same is hereby appropriated for the repairs and outfit of the steamship *Zavala*, also the sum of *twenty-five thousand dollars* for the outfit and provisioning of the Navy now in the employment of the government, and that the sums aforesaid be paid out of the first moneys in the Treasury, or at the disposition of the Executive.

SEC. 2. That the sum of *twenty-eight thousand two hundred and forty-one dollars* be, and the same is hereby appropriated for the pay of the officers, seamen and marines for services rendered, and due them on the first of July, 1842, also the sum of *twenty-nine thousand four hundred and twenty-eight dollars and fifty cents*, for pay of officers, seamen and marines of the Navy for the next succeeding six months from and after the 1st of July, 1842; and that in case any prizes should be made, or contributions levied by our Navy, then and in that case, the President is hereby authorized and required to apply the same, or so much thereof as may be requisite, to the payment of the above appropriations, in part payment of the officers, seamen and marines of the Navy.

...SEC. 4. That the officers of the Navy shall hereafter receive the following compensation: To a Post Captain, $200 per month; Commander, $120 per month; Lieutenants commanding, $100; Lieutenants, $80; Surgeons, $100; Assistant Surgeons, $80; Pursers, $80; Masters, $70; Midshipmen, $25; Warrant officers, $40; Secretary to Captain, $50; Clerks, $40; and marines and seamen shall be allowed pay by the law existing on the first Monday of September last.

Sec. 5. That should any prizes or contributions be made or received, the said officers, out of that portion of which government would be entitled, shall receive pro rata additional pay, until it shall amount to the compensation which they would have received under the laws in existence on the first Monday of September last.

Approved, July 23, 1842.
(See Laws of Texas, 6th Congr., special session, page 3.)

Sec. 1. *Be it enacted, &c.,* That the Secretary of War and Marine be, and he is hereby authorized, to receive proposals for keeping in ordinary the following vessels of the Navy of this Republic, to wit: The ship *Austin*, brigs *Wharton* and *Archer*, and schooner *San Bernard*.

Sec. 2. That any contract entered into for the keeping of said vessels in ordinary shall continue for one year from the date thereof, unless said vessels shall be sooner required for the public service; in which event the party of parties so contracting shall be paid according to contract, for the time he or they may have kept them.

Sec. 3. That the Secretary of War and Marine is hereby required to take bond, with good and sufficient security, from the party or parties contracting, conditioned for the faithful performance of such contract as may be entered into under the provisions of this act.

Sec. 4. That the sum of fifteen thousand dollars be, and the same is hereby appropriated to carry this act into effect.

Sec. 5. That the act approved 16th January, 1843, authorizing the sale of the Navy, be, and the same is hereby repealed.

Sec. 6. That if the Secretary of War and Marine shall not be able to make a contract, according to the provisions of this act, then, and in that case, he shall have power to apply the amount appropriated to the keeping of the Navy in ordinary, under his directions.

Approved, Feb. 3rd, 1844
(See Laws of Texas, 8th Congress, page 115.)

(Same Congress, and on the *next* page, 116.) SEC. 1. *Be it enacted, &c.,* That the sum of *sixteen thousand dollars,* be, and the same is hereby appropriated, for the part pay of the officers of the Navy.

Laws of Texas, ninth and last Congress, page 79. "For keeping the Navy in ordinary, *eight thousand dollars.*"

At the date of annexation, the law passed January 18, 1841, "to fix the Naval Establishment of the Republic," *was in full force and effect.* It is therefore *absurd to assert,* that "there were no officers in the Texas Navy at the annexation of that Republic to the United States."

The second section of the law, approved February 5, 1844, authorizing the vessels to be laid up in ordinary, clearly contemplates the probability of their being called into service, as is fully proven by the words, "unless said vessels shall be sooner required for the public service." No proposals were made, and the vessels were kept in ordinary under charge of officers and men of the Texas Navy—precisely as the same thing is done in the Navy of the United States. Placing a vessel of war in ordinary after a cruise does not in this government deprive the officers who were on board of her during her last cruise, of their commissions; neither did it by the laws of Texas, as would seem to be inferred by these *three* commanders in their two pamphlets addressed to your honorable body. If so, they would have been deprived of their commissions long since.

Fifth. It has been urged by some that the decision of the Supreme Court in the case of Brashear vs. Mason, (the Secretary of the Navy,) has some bearing on this subject. It will be seen by reference to this case that the only point involved in the controversy was the jurisdiction of the court to issue a mandamus to the Secretary of the Navy to do an executive act in a case where there was no specific appropriation out of which to pay the amount claimed.

(Brashear vs. Mason, 6 Howard's Reps, pages 101 & 102)

The Constitution provides that no money shall be drawn from the Treasury, but in consequence of appropriations made by law, (Art. 1, S. 9,) and it is declared by act of Congress, (3 St. U. S., p. 689, S. 2,) that all moneys appropriated for the use of the War and Navy Departments shall be drawn from the Treasury by warrants of the Secretary of Treasury, upon the requisitions of the Secretaries of these Departments, countersigned by the Second Comptroller.

And by the act of 1817 (3 St. U. S., 367, S. 8, 9,) it is made the duty of the Comptroller to countersign the warrants, only in cases when they shall be warranted by law. And all warrants drawn by the Secretary of the Treasury upon the Treasurer shall specify the particular appropriation to which the same shall be charged; and the moneys paid by virtue of such warrant, shall, in conformity thereto, be charged to such appropriation in the books kept by the Comptroller; and the sums appropriated for such branch of expenditure in the several Departments shall be solely applied to the object for which they are respectively appropriated, and no others. (2 St. U. S., 535 S. 1.)

Formerly, the moneys appropriated for the War and Navy Departments were placed in the treasury to the credit of the respective Secretaries. That practice has been changed, and all the moneys in the treasury are in to the credit, or in the custody of the treasurer; and can be drawn out, as we have seen, only on the warrant of the Secretary of the Treasury, countersigned by the Comptroller.

In the case of Mrs. Decatur vs. Paulding, (14 Pet. 497,) it was held by this court that a mandamus would not lie from the Circuit Court of the District to the Secretary of the Navy to compel him to pay to the plaintiff a sum of money claimed to be due as a pension under a resolution of Congress.

There was no question as to the amount due, if the plaintiff was properly entitled to the pension; and it was made to appear, in that case affirmatively on the application, that the pension fund was ample to satisfy the claim. The fund, also, was under the control of the Secretary, and the moneys payable on his own warrant.

Still the court refused to inquire into the merits of the claim of Mrs. D. to the pension, or whether it was rightfully withheld or not by the Secretary, on the ground that the court below had no jurisdiction over the case, and therefore the question was not properly before this court on the writ of error.

The court says that the duty required of the Secretary by the resolution was to be performed by him, as the head of one of the Executive Departments of the government, in the ordinary discharge of his official duties; that, in general, such duties, whether imposed by act of Congress or by resolution, are not mere ministerial duties; that the head of the Executive Department of the government, in the administration of the various and important concerns of his office, is continually required to exercise judgment and discretion.

And that the court could not, by mandamus, act directly upon the officer, and guide and control his judgment or discretion on matters committed to his care in the ordinary discharge of his official duties.

The court distinguished the case from Hendall vs. The United States (12 Pet. 524,) which was a mandamus to enforce the performance of a mere ministerial act, not involving, on the part of the officer, the exercise of any judgment or discretion.

The principle of the case of Mrs. Decatur is decisive of the present one. The facts here are much stronger to illustrate the inconvenience and unfitness of the remedy.

Besides the duty of inquiring into and ascertaining the rate of compensation that may be due the officers under the laws of Congress, no payment can be made unless there has been an appropriation for the purpose; and if made, it may have become already exhausted, or prior requisitions may have been issued sufficient to exhaust it.

The Secretary is obliged to inquire into the condition of the funds, and the claims already charged upon it, in order to ascertain if there is money enough to pay all the accruing demands; and if it be not enough how it shall be apportioned among the parties entitled to it.

These are important duties, calling for the exercise of judgment and discretion on the part of the officer, and in which the general creditors of the government, to the payment of whose demands the particular fund is applicable, are interested, as well as the government itself. At most, the Secretary is but a trustee of the fund, for the benefit of all those who have claims chargeable upon it, and, like other trustees, is bound to administer it with a view to the rights and interest of all concerned.

It will not do to say that the result of the proceeding by mandamus would show the title of the relator to his pay, the amount and whether there were any moneys in the treasury applicable to the demand for it; for, upon this ground, any creditor of the government would be enabled to enforce his claim against it, through the head of the proper department, by means of this writ;

and the proceedings would become as common, in the enforcement of demands upon the government, as the actions of assumpsit to enforce like demands against individuals.

For these reasons, we think the writ of mandamus would not lie in the case; and therefore, also, was properly refused by the court below, and that the judgment should be affirmed.

These *three* Commanders, in both their appeals to your honorable body, in reference to the bill reported by the Naval Committee, have been very urgent and industrious in *misrepresenting* my acts as Commanding Officer of the Texas Navy. All the orders cited by them as having been disobeyed by me *are purely conditional ones*. They, and other allegations made against me by the President of Texas, were the subject of a thorough investigation, which resulted in a *virtual acquittal* of every charge. The finding of the Court was quashed by the President of Texas, under whose administration the accusations were made and the trial held, by his endorsing on it the following:

The President disapproves the proceedings of the Court in toto, as he is assured by undoubted evidence of the guilt of the accused, in the case of E. W. Moore, Commander of the Navy.

(signed)   SAM. HOUSTON
7th Dec. 1844

The document, No. 3, a part of which these *three* Commanders append to their first pamphlet is dated June 27, 1845. It revives certain grave and serious charges against me, of which I had been acquitted more than six months previous. These are all personal issues, having no bearing upon the subjects proposed in the bill. Had they looked into this matter fully, they certainly would not have published any part of those documents. All the circumstances, however, easily to be reached, if not *within the actual knowledge* of these *three* Commanders, when they so unjustifiably published their version of them, which, I regret to say, has been industriously circulated among members of your honorable body, and of the Senate, evidently for the purpose of creating an undue prejudice against me, which I feel confident of explaining in the most satisfactory manner, whenever such explanation is necessary or proper. I do not consider this to be either the time or place, for the bill in question provides for incorporating into the Navy of the United States, *"only the surviving officers of the late Texas Navy, who were duly commissioned, and in the service of said Republic at the time of its annexation to the United States."*

Trusting in the high sense of justice of the Congress of the United States, and relying as we do, upon the merits of the case and what we believe to be our equitable claims under "liberal and just constructions of the terms of annexation," we hope that such action will be speedily taken by Congress as will enable us to participate in the union of our adopted country with this Confederacy.

All of which is respectfully submitted in behalf of the other memorialists and himself, by

> Your obedient servant,
> E. W. MOORE,
> *Captain Commanding late Texas Navy*
>
> WASHINGTON CITY
> February 4, 1851

[NOTE: The claims of the Texian officers were settled in 1857, with the surviving memorialists being awarded 5 years' pay at the going rate of U. S. Navy officers. Moore was among those who received the settlement and, in time, collected roughly $44,000 in claims against Texas.]